Nabil Derbel, Faouzi Derbel and Olfa Kanoun (Eds.)
Systems, Automation & Control

Advances in Systems, Signals and Devices

——

Edited by
Olfa Kanoun, University of Chemnitz, Germany

Volume 5

Systems, Automation & Control

—

Edited by
Nabil Derbel, Faouzi Derbel and Olfa Kanoun

DE GRUYTER
OLDENBOURG

Editors of this Volume

Prof. Dr.-Eng. Nabil Derbel
University of Sfax
Sfax National Engineering School
Control & Energy Management Laboratory
1173 BP, 3038 SFAX, Tunisia
n.derbel@enis.rnu.tn

Prof. Dr.-Ing. Faouzi Derbel
Leipzig University of Applied Sciences
Chair of Smart Diagnostic and Online Monitoring
Wächterstrasse 13
04107 Leipzig, Germany
faouzi.derbel@htwk-leipzig.de

Prof. Dr.-Ing. Olfa Kanoun
Technische Universität Chemnitz
Chair of Measurement and Sensor Technology
Reichenhainer Strasse 70
09126 Chemnitz
olfa.kanoun@etit.tu-chemnitz.de

ISBN 978-3-11-046821-2
e-ISBN (PDF) 978-3-11-047046-8
e-ISBN (EPUB) 978-3-11-046850-2
Set-ISBN 978-3-11-047047-5
ISSN 2365-7493

Library of Congress Cataloging-in-Publication Data
A CIP catalog record for this book has been applied for at the Library of Congress.

Bibliographic information published by the Deutsche Nationalbibliothek
The Deutsche Nationalbibliothek lists this publication in the Deutsche Nationalbibliografie;
detailed bibliographic data are available on the Internet at http://dnb.dnb.de.

© 2018 Walter de Gruyter GmbH, Berlin/Boston
Typesetting: Konvertus, Haarlem
Printing and binding: CPI books GmbH, Leck
♾ Printed on acid-free paper
Printed in Germany

www.degruyter.com

Preface of the Editors

The fifth volume of the Serie "Advances in Systems, Signals and Devices" (**ASSD**), is dedicated to fields related to "Systems, Automation and Control" (**SAC**). The scope of this issue encompasses all aspects of the research, development and applications of the science and technology in these fields.

Topics of this issue concerns: system design, system identification, biological and economical models & control, modern control theory, nonlinear observers, control and application of chaos, adaptive/non-adaptive backstepping control techniques, advances in linear control theory, system optimization, multivariable control, large scale and infinite dimension systems, nonlinear control, distributed control, predictive control, geometric control, adaptive control, optimal and stochastic control, robust control, neural control, fuzzy control, intelligent control systems, diagnostics, fault tolerant control, robotics and mechatronics, navigation, robotics and human-machine interaction, hierarchical and man-machine systems, etc.

Authors are encouraged to submit novel contributions which include results of research or experimental work discussing new developments in the field of systems, automation and control. The journal can be also addressed for editing special issues for novel developments in specific fields. Guest editors are encouraged to make proposals to the editor in chief of the corresponding main field.

The aims of this volume, in its own way, to promote an international scientific progress in the fields of systems, automation and control. It provides at the same time an opportunity to be informed about interesting results that have been reported during the international SSD conferences.

The Editors
Nabil Derbel, Faouzi Derbel and Olfa Kanoun

Advances in Systems, Signals and Devices

Series Editor:

Prof. Dr.-Ing. Olfa Kanoun
Technische Universität Chemnitz, Germany.
olfa.kanoun@etit.tu-chemnitz.de

Editors in Chief:

Systems, Automation & Control

Prof. Dr.-Eng. Nabil Derbel
ENIS, University of Sfax, Tunisia
n.derbel@enis.rnu.tn

Power Systems & Smart Energies

Prof. Dr.-Ing. Faouzi Derbel
Leipzig Univ. of Applied Sciences, Germany
faouzi.derbel@htwk-leipzig.de

Communication, Signal Processing & Information Technology

Prof. Dr.-Ing. Faouzi Derbel
Leipzig Univ. of Applied Sciences, Germany
faouzi.derbel@htwk-leipzig.de

Sensors, Circuits & Instrumentation Systems

Prof. Dr.-Ing. Olfa Kanoun
Technische Universität Chemnitz, Germany
olfa.kanoun@etit.tu-chemnitz.de

Communication, Signal Processing & Information Technology

Til Aach, Achen University, Germany
Kasim Al-Aubidy, Philadelphia Univ., Amman, Jordan
Adel Alimi, Engineering School of Sfax, Tunisia
Najoua Benamara, Engineering School of Sousse, Tunisia
Ridha Bouallegue, Engineering School of Sousse, Tunisia
Dominique Dallet, ENSEIRB, Bordeaux, France
Mohamed Deriche, King Fahd University, Saudi Arabia
Khalifa Djemal, Université d'Evry, Val d'Essonne, France
Daniela Dragomirescu, LAAS, CNRS, Toulouse, France
Khalil Drira, LAAS, CNRS, Toulouse, France
Noureddine Ellouze, Engineering School of Tunis, Tunisia
Faouzi Ghorbel, ENSI, Tunis, Tunisia
Karl Holger, University of Paderborn, Germany
Berthold Lankl, Univ. Bundeswehr, München, Germany
George Moschytz, ETH Zürich, Switzerland
Radu Popescu-Zeletin, Fraunhofer Inst. Fokus, Berlin, Germany
Basel Solimane, ENST, Bretagne, France
Philippe Vanheeghe, Ecole Centrale de Lille France

Sensors, Circuits & Instrumentation Systems

Ali Boukabache, Univ. Paul, Sabatier, Toulouse, France
Georg Brasseur, Graz University of Technology, Austria
Serge Demidenko, Monash University, Selangor, Malaysia
Gerhard Fischerauer, Universität Bayreuth, Germany
Patrick Garda, Univ. Pierre & Marie Curie, Paris, France
P. M. B. Silva Girão, Inst. Superior Técnico, Lisboa, Portugal
Voicu Groza, University of Ottawa, Ottawa, Canada
Volker Hans, University of Essen, Germany
Aimé Lay Ekuakille, Università degli Studi di Lecce, Italy
Mourad Loulou, Engineering School of Sfax, Tunisia
Mohamed Masmoudi, Engineering School of Sfax, Tunisia
Subha Mukhopadhyay, Massey University Turitea, New Zealand
Fernando Puente León, Technical Univ. of München, Germany
Leonard Reindl, Inst. Mikrosystemtec., Freiburg Germany
Pavel Ripka, Tech. Univ. Praha, Czech Republic
Abdulmotaleb El Saddik, SITE, Univ. Ottawa, Ontario, Canada
Gordon Silverman, Manhattan College Riverdale, NY, USA
Rached Tourki, Faculty of Sciences, Monastir, Tunisia
Bernhard Zagar, Johannes Kepler Univ. of Linz, Austria

Advances in Systems, Signals and Devices

Volume 1
N. Derbel (Ed.)
Systems, Automation, and Control, 2016
ISBN 978-3-11-044376-9, e-ISBN 978-3-11-044843-6, e-ISBN (EPUB) 978-3-11-044627-2, Set-ISBN 978-3-11-044844-3

Volume 2
O. Kanoun, F. Derbel, N. Derbel (Eds.)
Sensors, Circuits and Instrumentation Systems, 2016
ISBN 978-3-11-046819-9, e-ISBN 978-3-11-047044-4, e-ISBN (EPUB) 978-3-11-046849-6, Set-ISBN 978-3-11-047045-1

Volume 3
F. Derbel, N. Derbel, O. Kanoun (Eds.)
Power Systems & Smart Energies, 2016
ISBN 978-3-11-044615-9, e-ISBN 978-3-11-044841-2, e-ISBN (EPUB) 978-3-11-044628-9, Set-ISBN 978-3-11-044842-9

Volume 4
F. Derbel, N. Derbel, O. Kanoun (Eds.)
Communication, Signal Processing & Information Technology, 2016
ISBN 978-3-11-044616-6, e-ISBN 978-3-11-044839-9, e-ISBN (EPUB) 978-3-11-043618-1, Set-ISBN 978-3-11-044840-5

Contents

S. Ghachem, K. Benothman and M. Benrejeb

Nonlinear Progressive Accommodation of Actuator Fault

Abstract: In this paper, a method of actuator fault tolerant control for a class of nonlinear systems is proposed. It concerns the problem of nonlinear progressive accommodation to actuator failure. This strategy is based on the optimal nonlinear controller which obtained by solving the Generalized Hamilton–Jacobi–Bellman Equation, and its objective is to maintain the system closed loop stability when an actuator fault appear. Two examples are given to illustrate this approach.

Keywords: Optimal Control, Nonlinear Systems, Nonlinear Progressive Accommodation, Fault Tolerant Control, Actuator Fault.

1 Introduction

Modern technological systems rely on sophisticated control functions to meet increased performance requirements. For such systems, Fault Tolerant Control (FTC) needs to be developed. A FTC is a control that can accommodate system component faults and it is able to maintain stability and acceptable degree of performance with respect to nominal system operation [1] and accept some graceful performance degradation [2] not only when the system is fault-free but also when there are component malfunctions.

FTC can be classified into passive and active. A Passive FTC (PFTC) can tolerate a predefined set of faults while accomplishing its mission satisfactory without the need for control reconfiguration. Active FTC (AFTC), on the other hand, relies on a Fault Detection and Identification (FDI) process to monitor system performance, and to detect and isolate faults in the system. Accordingly, the control law is reconfigured on-line [3, 4].

Indeed, in the literature, a conventional strategy to solve a nonlinear reconfigurable control problem consists in designing a linear approximation of the model around operating points. Recent papers such as multiple model [5] sliding modes [6] and internal model control [7] have been presented, and in order to handle nonlinear systems beyond using a linearized approximation, reconfigurable control methods have been proposed using backstepping [8] and nonlinear regulator [9].

S. Ghachem, K. Benothman and M. Benrejeb: Research Laboratory on Automatic (LARA), University of Tunis El Manar, Tunis Engineering School, BP 37, Belvedere 1002 Tunis, Tunisia., email: xxx@xxx

De Gruyter Oldenbourg, ASSD – Advances in Systems, Signals and Devices, Volume 5, 2018, pp. 1–16.
DOI 10.1515/9783110470468-001

Moreover, few papers concern the delays associated with computation times [10, 11, 12]. The former introduced the concept of progressive accommodation whose the objective is to minimize the effect of the accommodation delay. To this end, the reconfigurable control design method is based on a linear quadratic approach and it's named Linear Progressive Accommodation (LPA).

But, the limitation of this method is presented in our last works [13, 14, 15], when an analysis of the closed loop system stabilization during the fault occurrence with the use of the domain of attraction and the linear approximation validity domain, proof the divergence of the Linear Progressive Accommodation (LPA) when the actuator fault occurs near the boundary of the validity domain of the linearization. For solving this problem, tow approaches are presented in the last work [14, 15, 16].

In this paper, the authors propose an approach of fault tolerant control for a class of affine nonlinear system. This strategy is based on the optimal nonlinear controller which obtained from an iterative algorithm by solving the Generalized Hamilton–Jacobi–Bellman Equation (HJBE) with the Successive Galerkin Approximation (SGA) [17], and its objective to maintain the system closed loop stability when an actuator fault appear and to minimize the effect of the accommodation delay. This method is called Nonlinear Progressive Accommodation (NLPA).

The present paper is organized as follows: in section 2, the class of affine nonlinear systems is introduced and a necessary background is provided on the main idea of the actuator fault accommodation and optimal regulation problem. Section 3 presents the proposed approach of nonlinear progressive accommodation. In section 4, simulation studies have been conducted in two examples to illustrate the proposed approach.

2 Preliminaries and Motivation

In the present work, affine nonlinear continuous-time dynamic systems are considered with a state-space representation:

$$\dot{x} = f(x) + Bu \tag{1}$$

$$y = h(x) \tag{2}$$

where $x \in \mathbb{R}^n$ is the vector of state variables, $u \in \mathbb{R}^m$ is the control vector and $y \in \mathbb{R}^l$ is the output vector. f and h are smooth functions with $f(0) = 0$. B is a constant matrix of dimension $(n \times m)$. The infinite-time horizon nonlinear regulation problem is defined with the following quadratic performance index in u:

$$V(x) = \min_u \int_0^\infty (x^T Q(x)x + u^T R(x)u)dt \tag{3}$$

in which $Q(x) \geq 0$ and $R(x) > 0$ for all x. Moreover, it is assumed that Q and R are sufficiently smooth so that the value function $V(x)$ is continuously differentiable. In this case, the Hamilton–Jacobi Equation (HJE) is quadratic in $\frac{\partial V}{\partial x}(x)$ such that:

$$\frac{\partial V}{\partial x}(x)f(x) - \frac{1}{4}\frac{\partial V}{\partial x}(x)BR^{-1}B^T\frac{\partial V^T}{\partial x}(x) + x^TQx = 0 \tag{4}$$

and the optimal feedback control can be designed from:

$$u = -\frac{1}{2}R^{-1}(x)B^T\frac{\partial V^T}{\partial x}(x) \tag{5}$$

In this paper, we consider that one (or several) actuator fault(s) occur at time t_f. The system (1) can be described by:

$$\dot{x} = f(x) + B_\theta(u), \tag{6}$$

where:

$$B_\theta(u) = \begin{cases} Bu, & t \in [0, t_f[\\ \beta_f(u, \theta), & t \in [t_f, \infty[\end{cases} \tag{7}$$

The function $\beta_f(u, \theta)$ and the parameter θ represent the contribution of the faulty actuator. The complex structure of the system (1) introduces difficulties in solving the optimal control problem. The calculation of an optimal nonlinear state feedback for nonlinear systems requires the development of numerical algorithms, because the optimization problem needs a resolution of the Hamilton–Jacobi equation. As mentioned in [11], in the FTC problem, one has to consider four time periods in order to analyze the system behavior under actuator fault:

- $t \in [0, t_f[$: nominal system and control u_n.
- $t \in [t_f, t_{fdi}[$: faulty system under the nominal control u_n and FDI algorithm in process for fault detection, isolation and estimation.
- $t \in [t_{fdi}, t_{ftc}[$: faulty system under the nominal control u_n and the fault is detected, isolated and estimated.
- $t \in [t_{ftc}, \infty[$: faulty system under the accommodated control u_f.

These four time periods are presented in Fig. 1.

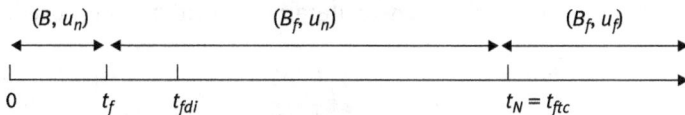

Fig. 1. Description of the fault tolerant control strategy.

In practical applications, even if the diagnosis is perfect that is not realistic, the system control is inappropriate on the interval $[t_f, t_{ftc}[$ since the faulty system is controlled by u_n. The progressive accommodation aims at minimizing the interval $[t_{fdi}, t_{ftc}[$.

Therefore, thanks to an online control computation, the authors propose an improvement of the closed loop behaviour of the fault system in a nonlinear context.

3 Proposed Approach: Nonlinear Progressive Accommodation (NLPA)

In practice, an actuator fault in a controlled system generates changes in inputs/outputs signals and in the parameters of the differential system which describes the dynamics. The design of a passive fault tolerant controller is sufficient to ensure degraded dynamic performances when the changes in the parameters and signals are small. When the effects of the fault are significant, the global stability of the system may not be ensured, therefore the stabilization of the dynamic system with a fixed controller may be impossible.

In this paper, the authors consider an actuator fault occurrence under the constraint that the faulty actuator can?t be switched-off and replaced. This last strategy is usually called system reconfiguration. In this section, the authors focus their attention on the fault accommodation in a nonlinear context. They first refer to a fault tolerant control designed beforehand when failure is identified and secondly to an on-line accommodation scheme.

3.1 Nonlinear Progressive Accommodation

In the nonlinear case, the infinite-time horizon nonlinear optimal control problem (1), (3), is characterized in terms of Hamilton–Jacobi Equation (4). The difficulty with finding the optimal control is that the HJB equation represents a nonlinear partial differential equation, which is difficult, and often impossible, to solve analytically. The complexity of the HJE prevents any solution excepted in some very simple systems.

In order to make real-time implementation possible, one has to avoid solving any partial differential equation. With application to online progressive accommodation and in order to design a suboptimal control design, an iterative algorithm is proposed:

$$u_i(x) = \begin{cases} u_0(x), & i = 0 \\ -\frac{1}{2}R^{-1}B^T \frac{\partial V_{i-1}^T}{\partial x}(x), & i > 0 \end{cases} \tag{8}$$

where V_{i-1} is the performance index of u_{i-1} as calculated from the solution of the Generalized Hamilton–Jacobi–Bellman (GHJB) equation:

$$\frac{\partial V_{i-1}}{\partial x}(f(x) + Bu_{i-1}) + u_{i-1}^T R^{-1} u_{i-1} + x^T Qx = 0 \tag{9}$$

if the u_0 is stabilizing control, then u_i will be stabilizing for all $i \geq 0$, $V_i \to V$ and $u_i \to u$ as $i \to \infty$.

To solving the Generalized Hamilton–Jacobi–Bellman (GHJB) equation, an alternative is to utilize the Successive Galerkin Approximation (SGA) [17, 18, 19].

3.2 Successive Galerkin Approximation (SGA)

The Successive Galerkin Approximation (SGA) algorithm is based on a simultaneous approximation in policy space and a Galerkin spectral approximation. The algorithm approximates the optimal control by a truncated series of a complete set of basis functions. If the basis functions are formed by taking the tensor product of a complete set of one-dimensional functions then the number of basis functions required for an Mth order approximation grows exponentially in the size of the state space. In addition, the coefficients associated with the basis functions are computed via a projection operator on a subset of : requiring the computations of multidimensional integrals. The conditions guaranteeing the convergence of the algorithm are reported in [17, 20, 21].

The Galerkin series approximation, V_i^N, of equation (9), in a series solution in terms of N elements of a complete set of basis functions $\{\phi_j(x)\}_1^\infty$:

$$V_i^N = \sum_{j=0}^N c_i^j \phi_j(x) = \phi_N^T C_i^N \tag{10}$$

where $\phi_N(x) = (\phi_1(x), \dots, \phi_N(x))^T$ and $C_i^N = (c_i^1, \dots, c_i^N)^T$. By substituting V_i^N into equations (8) and (9) we achieve the following result.

The coefficients, C_i^N, of the Galerkin approximation of the GHJB equation can be found by solving the following equations:
a) if $i = 0$,

$$(b_2 + b_4)C_0^N + (b_1 + b_5) = 0 \tag{11}$$

b) if $i \geq 0$,

$$\left(b_2 - \frac{1}{2}b_3[C_{i-1}^N \otimes I_N]\right)C_i^N + \left(b_1 + \frac{1}{4}b_3[C_{i-1}^N \otimes C_{i-1}^N]\right) \tag{12}$$

where,

$$b_1 = \int_\Omega x^T Q x \Phi_N(x) dx \tag{13}$$

$$b_2 = \sum_{j=1}^{n} \int_\Omega f_j(x) \Phi_N(x) \frac{\partial \Phi_N^T}{\partial x_j} dx \tag{14}$$

$$b_3 = \sum_{j=1}^{n} \sum_{k=1}^{n} \int_\Omega B_j R^{-1} B_k^T \left(\frac{\partial \Phi_N^T}{\partial x_k} \otimes \Phi_N(x) \frac{\partial \Phi_N^T}{\partial x_j} \right) dx \tag{15}$$

$$b_4 = \sum_{j=1}^{n} \int_\Omega (Bu_0)_j \Phi_N(x) \frac{\partial \Phi_N^T}{\partial x_j} dx \tag{16}$$

$$b_5 = \int_\Omega u_0^T R u_0 \Phi_N(x) dx \tag{17}$$

and I_N is the N dimensional identity matrix. $\otimes$ is the standard Kronecker product of two matrices.

4 Illustrative Exemple 1

Consider an affine nonlinear continuous-time dynamic system modeled by:

$$\dot{x} = \begin{pmatrix} -x_1^3 - x_2 \\ x_1 + x_2 \end{pmatrix} + \begin{pmatrix} 0 \\ 1 \end{pmatrix} u \tag{18}$$

$$y = x \tag{19}$$

where $x \in \mathbb{R}^2$ is the state vector, $u \in \mathbb{R}$ is the control vector and $y \in \mathbb{R}$ is the output vector.

With $f(x) = \begin{pmatrix} -x_1^3 - x_2 \\ x_1 + x_2 \end{pmatrix}$ and $B = \begin{pmatrix} 0 \\ 1 \end{pmatrix}$ is a constant matrix.

The following problem is first to define an optimal control u_n with respect to a quadratic performance index (3), in nominal conditions given an initial value of the state x_0. Secondly, a fault tolerant control u_f must be synthesized given an acceptable actuator fault.

4.1 Nonlinear Optimal Control in Nominal Conditions

The quadratic performance index (3), with $Q(x) = I_2 = \begin{pmatrix} 1 & 0 \\ 0 & 1 \end{pmatrix}$ and $R(x) = 1$, becomes:

$$V(x) = \min_u \int_0^\infty (x^T I_2 x + u^T u)dt \qquad (20)$$

where $x \in \mathbb{R}^2$ is the vector of state variables, $u \in \mathbb{R}$ is the control vector.

The nonlinear nominal optimal control which minimizes the performance index (20) and satisfies Bellman's Principle is given by:

$$u_n(x) = -\frac{1}{2} B^T \frac{\partial V^T}{\partial x}(x) \qquad (21)$$

where $V(x)$, the minimal performance index is found by solving for the stabilizing solution of the Hamilton–Jacobi–Bellman (HJB) equation:

$$\frac{\partial V}{\partial x}(x)f(x) - \frac{1}{4}\frac{\partial V}{\partial x}(x)BB^T\frac{\partial V^T}{\partial x}(x) + x^T I_2 x = 0 \qquad (22)$$

The nonlinear nominal optimal control is:

$$u_n(x) = 3x_1^5 - 1.3522x_1^3 + 3x_1^2 x_2 + 0.4142x_1 - 2.3522x_2 \qquad (23)$$

As an illustration, an initial condition $x_0 = (0.1 \ 0.1)^T$ is given. Figures 2 and 3 illustrate the output $x_1(t)$ and $x_2(t)$ with the nonlinear nominal optimal control.

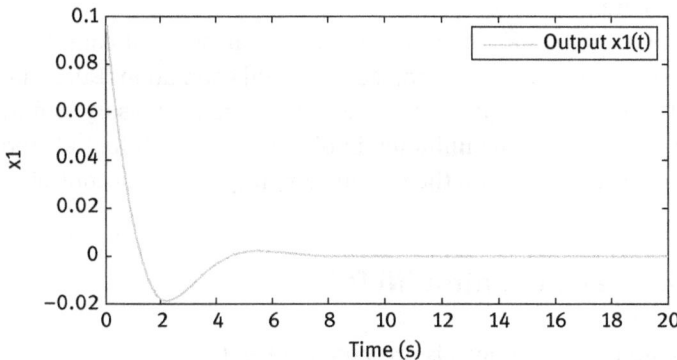

Fig. 2. Output $x_1(t)$ with nonlinear nominal optimal control.

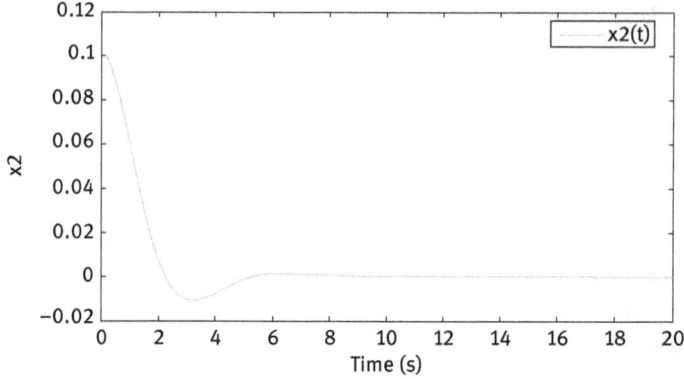

Fig. 3. Output $x_2(t)$ with nonlinear nominal optimal control.

4.2 Nonlinear Faulty System

An actuator fault occurs at the time $t_f = 0.2s$. According to the definition (7), the nonlinear faulty system is described by:

$$\begin{cases} \dot{x} = f(x) + Bu, & t \in [0, t_f[\\ \dot{x} = f(x) + B_f u, & t \in [t_f, +\infty[\end{cases} \tag{24}$$

where $B_f = 0.4B = (0 \ \ 0.4)^T$, then become:

$$\begin{cases} \dot{x} = \begin{pmatrix} -x_1^3 - x_2 \\ x_1 + x_2 \end{pmatrix} + \begin{pmatrix} 0 \\ 1 \end{pmatrix} u, & t \in [0, t_f[\\ \dot{x} = \begin{pmatrix} -x_1^3 - x_2 \\ x_1 + x_2 \end{pmatrix} + \begin{pmatrix} 0 \\ 0.4 \end{pmatrix} u, & t \in [t_f, +\infty[\end{cases} \tag{25}$$

u_f is the nonlinear fault tolerant control.

Before finding u_f, the faulty system is under the nonlinear nominal control u_n and FDI algorithm in process for fault detection, isolation and estimation. But if the nonlinear nominal control u_n can't stabilize the faulty system, it's necessary to find the nonlinear fault tolerant control u_f in minimum time. Figures 4 and 5 illustrate the output $x_1(t)$ and $x_2(t)$ in faulty system with the nonlinear nominal optimal control.

4.3 Nonlinear Fault Tolerant Control (NFTC)

The nonlinear fault tolerant control strategy is presented in Fig. 1.

Let consider the sample computation time $t_e = 0.1s$ and one supposes that the time delay for the fault diagnosis $t_{fdi} - t_f = t_e$ is equal to t_e.

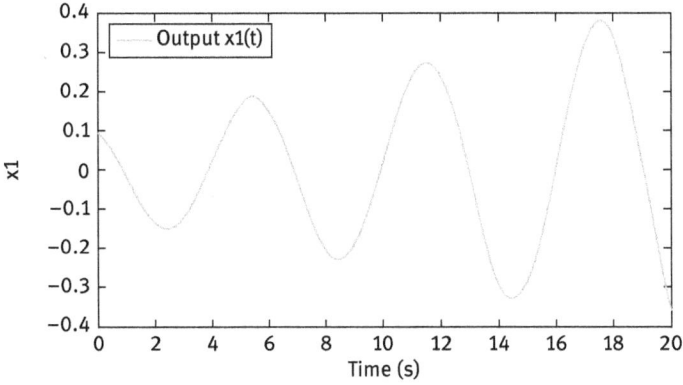

Fig. 4. Output $x_1(t)$ in faulty system with nonlinear nominal optimal control.

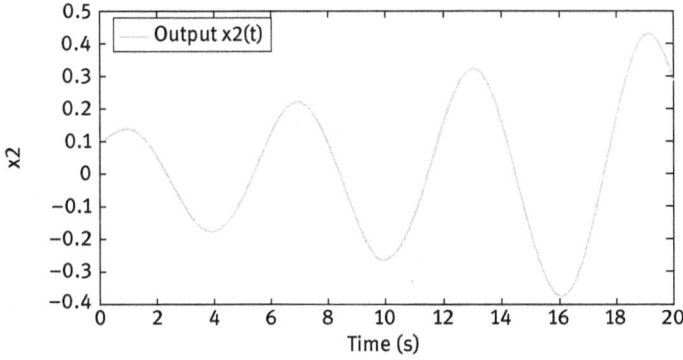

Fig. 5. Output $x_2(t)$ in faulty system with nonlinear nominal optimal control.

The nonlinear fault tolerant control u_f is given by:

$$u_f(x) = -\frac{1}{2}B_f^T \frac{\partial V_f^T}{\partial x}(x) \tag{26}$$

where $V_f(x)$, the minimal performance index is found by solving for the stabilizing solution of the Hamilton–Jacobi–Bellman (HJB) equation:

$$\frac{\partial V_f}{\partial x}(x)f(x) - \frac{1}{4}\frac{\partial V_f}{\partial x}(x)B_fB_f^T \frac{\partial V_f^T}{\partial x}(x) + x^T I_2 x = 0 \tag{27}$$

The nonlinear fault tolerant control is:

$$u_f(x) = 12x_1^5 - 5.4088x_1^3 + 12x_1^2 x_2 + 1.6568x_1 - 9.4088x_2 \tag{28}$$

With the same initial condition $x_0 = (0.1 \ \ 0.1)^T$. Figures 6 and 7 illustrate the output $x_1(t)$ and $x_2(t)$ with the nonlinear fault tolerant control.

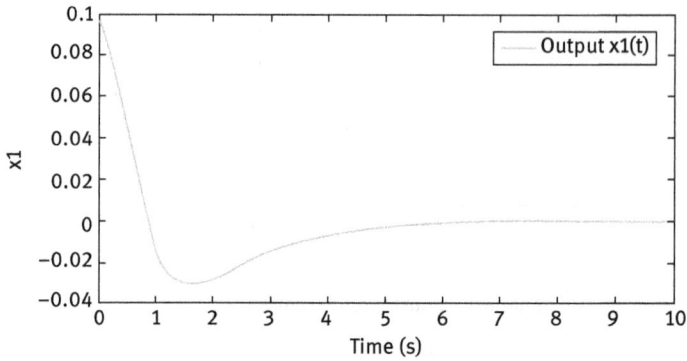

Fig. 6. Output $x_1(t)$ in faulty system with nonlinear fault tolerant control.

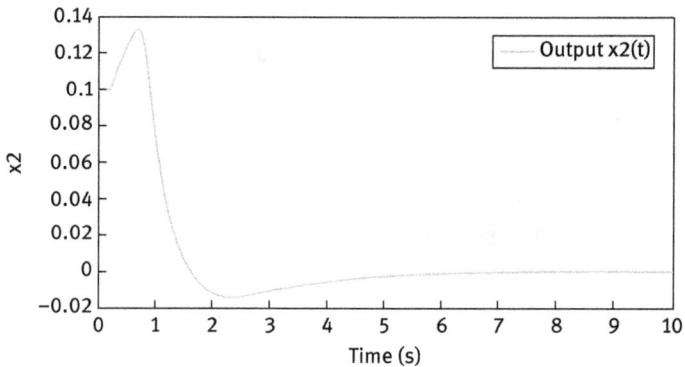

Fig. 7. Output $x_2(t)$ in faulty system with nonlinear fault tolerant control.

4.4 Nonlinear Progressive Accommodation (NLPA)

In the NLPA problem, one has to consider six time periods in order to analyze the system behavior under actuator fault:

- $t \in [0, t_f[$: nominal system and control u_n.
- $t \in [t_f, t_{fdi}[$: faulty system under the nominal control u_n and FDI algorithm in process for fault detection, isolation and estimation.

- $t \in [t_{fdi}, t_0[$: faulty system under the nominal control u_n and the instant t_0 represent the initialization progressive accommodation algorithm.
- $t \in [t_0, t_1[$: faulty system under the control u_0 which represent the initialization progressive accommodation algorithm (in this example $u_0 = u_n$).
- $t \in [t_i, t_{i+1}[$: faulty system under the control u_i which represent the ith iteration progressive accommodation algorithm.
- $t \in [t_{ftc}, +\infty[$: faulty system under the control u_N which represent the Nth iteration progressive accommodation algorithm, and $u_n = u_f$ (u_f is the accommodated control).

These six time periods are presented in Fig. 8.

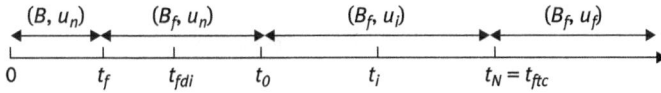

Fig. 8. Description of progressive accommodation strategy.

With the same initial condition $x_0 = (0.1 \quad 0.1)^T$. Figures 9 and 10 illustrate the output $x_1(t)$ and $x_2(t)$ with the nonlinear fault tolerant control.

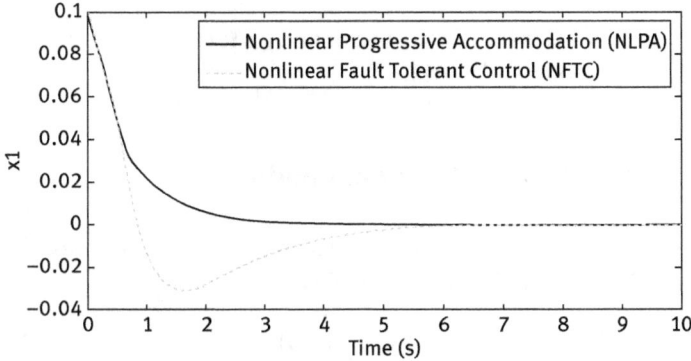

Fig. 9. Output $x_1(t)$ in faulty system with nonlinear progressive accommodation.

The plots show the improvement of the nonlinear progressive accommodation method for the actuator fault best than the method which utilize the nonlinear fault tolerant control.

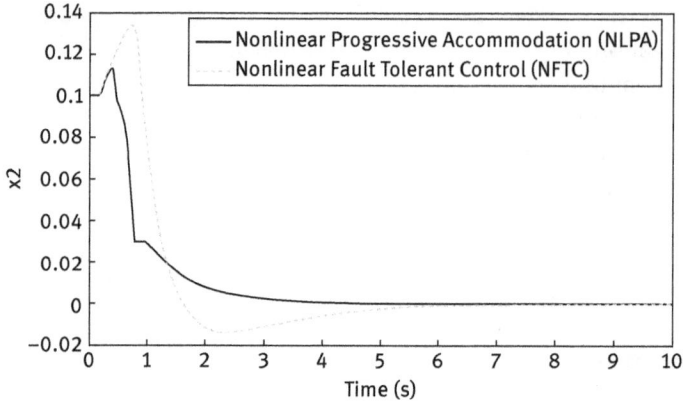

Fig. 10. Output $x_2(t)$ in faulty system with nonlinear progressive accommodationl.

5 Illustrative Exemple 2

Consider an affine nonlinear continuous-time dynamic system:

$$\dot{x} = x + x^2 + 2u \tag{29}$$

$$y = x \tag{30}$$

where $x \in \mathbb{R}$ is the state vector, $u \in \mathbb{R}$ is the control vector and $y \in \mathbb{R}$ is the output vector. With $f(x) = x + x^2$ and $B = 2$ is a constant matrix.

5.1 Nonlinear Optimal Control in Nominal Conditions

With $Q = R = 1$, the nonlinear nominal optimal control which minimizes the performance index (31) and satisfies Bellman's Principle is:

$$u_n(x) = -0.5x(x + 1 + \sqrt{x^2 + 2x + 5}) \tag{31}$$

As an illustration, an initial condition $x_0 = 0.1$ is given. Figure 11 illustrate the output $x(t)$ with the nonlinear nominal optimal control.

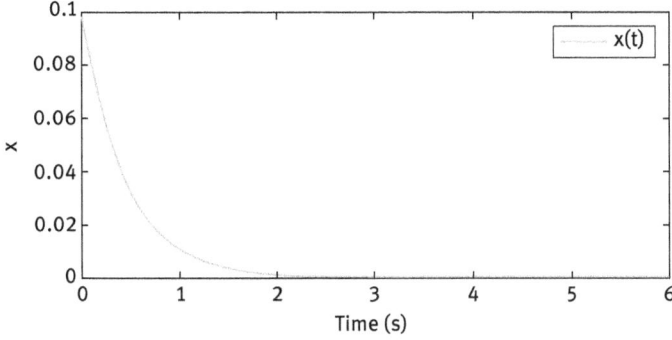

Fig. 11. Output $x(t)$ with nonlinear nominal optimal control.

5.2 Nonlinear Faulty System

An actuator fault occurs at the time $t_f = 0.2s$. According to the definition (7), the nonlinear faulty system is described by:

$$\begin{cases} \dot{x} = x + x^2 + 2u, & t \in [0, t_f[\\ \dot{x} = x + x^2 + 0.8u, & t \in [t_f, +\infty[\end{cases} \tag{32}$$

where $B_f = 0.8$.

5.3 Nonlinear Fault Tolerant Control (NFTC)

Let consider the sample computation time $t_e = 0.1s$ and one supposes that the time delay for the fault diagnosis $(t_{fdi} - t_f)$ is equal to t_e. The nonlinear fault tolerant control u_f is:

$$u_f(x) = -1.25x(x + 1 + \sqrt{x^2 + 2x + 1.64}) \tag{33}$$

with the same initial condition $x_0 = 0.1$ is given. Figure 12 illustrate the output $x(t)$ with the nonlinear fault tolerant control.

5.4 Nonlinear Progressive Accommodation (NLPA)

With the same initial condition $x_0 = 0.1$. Figure 13 illustrate the output $x(t)$ with the nonlinear progressive accommodation.

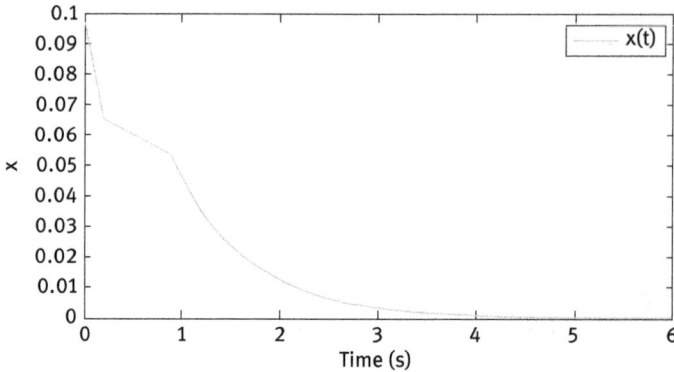

Fig. 12. Output $x(t)$ with nonlinear fault tolerant control.

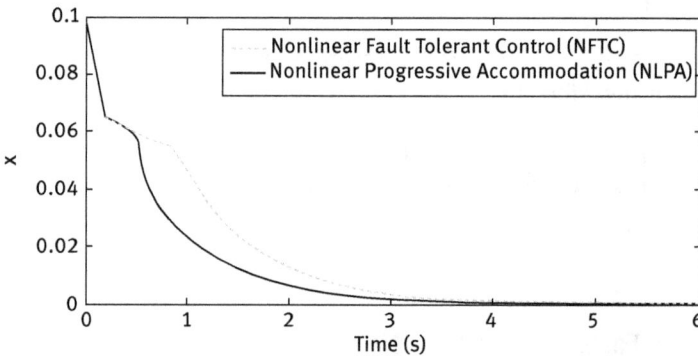

Fig. 13. Output $x(t)$ with nonlinear progressive accommodation.

The plots show the improvement of the nonlinear progressive accommodation method for the actuator fault best than the method which utilize the nonlinear fault tolerant control.

6 Conclusion

This paper emphasizes the importance of improving the system behavior during the fault accommodation delay. In the case of affine continuous-time dynamic nonlinear system and thanks to an iterative algorithm taking to account the nonlinearity in the active fault tolerant control synthesis, the nonlinear progressive accommodation scheme is proved to be efficient and best than classic method and shows that it can minimize the large transients which may occur when the faulty system is still controlled by the nominal control law. This strategy is based on the optimal nonlinear controller which obtained by solving the Generalized Hamilton–Jacobi–Bellman Equation, and its objective is to maintain the system closed loop stability when an actuator fault appear. Future research will investigate more algorithms for solving the Generalized Hamilton–Jacobi–Bellman Equation.

Bibliography

[1]   M. Blanke, M. Kinnaert, J. Lunze, and M. Staroswiecki. *Diagnosis and Fault Tolerant Control*. Springer Verlag, Berlin, 2006.
[2]   J. Jiang, and Y. Zhang. Accepting performance degradation in fault tolerant control system design. *IEEE Trans. on Control Systems Technology*, 14(2):284–292, 2006.
[3]   J. Chen, and R.J. Patton. *Robust model-based fault diagnosis for dynamic systems*. Kluwer, London, 1999.
[4]   H. Rauch. Autonomous control reconfigurations. *IEEE Control Systems Magazine*, 15(6):37–48, 1995.
[5]   D. Theillol, D. Sauter, and J.C. Ponsart. A multiple model based approach for fault tolerant control nonlinear systems. 5th *IFAC symposium on fault detection, supervision and safety for technical processes*,:151–156, Washington, 2003.
[6]   R.A. Hess, and S.R. Wells. Sliding mode control applied to reconfigurable flight control design. *J. of Guidance, Control and Dynamics*, 26(3):452–462, 2003.
[7]   D. Kim, D. Lee, and K.C. Veluvolu. Accommodation of actuator fault using local diagnosis and IMC-PID. *Int. J. of Control, Automation and Systems*, 12(6):1257–1265, 2014.
[8]   X. Zhang, and M.M. Polycarpou. Integrated design of fault diagnosis and accommodation schemes for a class of nonlinear systems. 44th *IEEE Conf. Decision and Control*,:1448–1453, Orlando, 2001.
[9]   G. Bajpai, B.C. Chang, and H.G. Kwatny. Design of fault tolerant systems for actuator failures in nonlinear systems. *IEEE American Control Conf.*,:3618–3623, Anchorage, 2002.
[10] M. Staroswiecki, H. Yang, and B. Jiang. Progressive accommodation of parametric faults in linear quadratic control. *Automatica*, 43(12):2070–2076, 2007.
[11] M. Staroswiecki. Progressive accommodation of actuator faults in the linear quadratic control problem. 43th *IEEE Conf. Decision and Control*,:5234–5241, Bahamas, 2004.
[12] X. Zhang, T. Parisini, and M.M. Polycarpou. Adaptive fault tolerant control of nonlinear uncertain systems: an information based diagnostic approach. *IEEE Trans. on Automatic Control*, 49(8):1259–1274, 2004.
[13] T. Raharijaona, R. Outbib, M. Ouladsine, and S. Ghachem. Comments on Actuator Fault Accommodation. 7th *IFAC Symposium on Fault Detection, Supervision and Safety of Technical Processes*,:1611–1616, Barcelona, 2009.
[14] S. Ghachem, K. Ben Othman, and M. Benrejeb. Nonlinear approach of actuator fault accommodation. *International Journal of Computer Science Issues*, 8(4):290–297, 2011.
[15] S. Ghachem, K. Ben Othman, and M. Benrejeb. Actuator Fault Accommodation with Optimal Control for a Class of Nonlinear Systems. *International Review of Automatic Control*, 5(2):274–283, 2012.
[16] S. Ghachem, K. Ben Othman, and M. Benrejeb. Nonlinear Progressive Accommodation of Actuator Fault with Optimal Control. 10th *IEE Int. Multi-Conf. on Systems, Signals and Devices*, Hammamet, 2013.

[17] R. Beard, G. Saridis, and J. Wen. Galerkin approximation of the generalized Hamilton Jacobi Bellman equation. *Automatica*, 33(12):2159–2177, 1997.
[18] J. Lawton, R. W. Beard and T. McLain. Successive Galerkin Approximation Of Nonlinear Optimal Attitude Control. *IEEE American Control Conf.*,:4373–4377, California, 1999.
[19] J. Lawton, and R. W. Beard. Numerically efficient approximations to the Hamilton–Jacobi–Bellman equation. *IEEE American Control Conf.*,:195–199, Philadelphia, 1998.
[20] R. Beard, G. Saridis, and J. Wen. Improving the performance of stabilizing control for nonlinear systems. *Control Systems Magazine*,:27–35, 1996.
[21] R. Beard, G. Saridis, and J. Wen. Approximate solutions to the time-invariant Hamilton–Jacobi–Bellman equation. *Journal of Optimization Theory and Applications*, 96(3):589–626, 1998.

Biographies

Sahbi Ghachem was born in Tunisia in 1976. He obtained the Engineer degree in electro-Mechanical engineering from the Engineer National School of Sfax in 2003. He obtained the Master degree in Automatic and industrial Maintenance from the Engineer National School of Monastir in 2005. His research is related to Reliability, Fault Diagnosis and Fault tolerant control System.

Kamel Benothman was born in Tunisia in 1958. He obtained the Engineer degree in Mechanical and Energetic engineering from the University of Valenciennes in 1981. He obtained the PhD degree in Automatic and Signal Processing from the University of Valenciennes in 1984 and the HDR degree from the Engineer National School of Tunis in 2008. He is currently a professor in the Superior Institute of Science and Energy Technologies of Gafsa. His research is related to Reliability, fuzzy systems and Diagnosis of complex systems.

Mohamed Benrejeb was born in Tunisia in 1950. He obtained the Engineer degree from the Central School of Lille in 1973, The Master degree of Automatic from the Central School of Lille in 1974, the Doctor-Engineer degree in Automatic from the Central School of Lille in 1976. He obtained the PhD degree in Physical Sciences from the University of Science and Technology of Lille in 1980. He is a Professor in the Engineer National School of Tunis since 1985 and in Central School of Lille since 2003. His research focuses on the analysis and synthesis of complex continuous and discrete systems, based on conventional and unconventional approaches, and more recently on problem solving scheduling of workshops and study event systems discrete. This work has already led to some fifty publications in international journals and a book on artificial neural networks and applications.

M. Bennehar, A. Chemori, F. Pierrot and S. Krut

Control of Redundantly Actuated PKMs for Closed-Shape Trajectories Tracking with Real-Time Experiments

Abstract: This paper deals with closed-shape geometric reference trajectories generation and dynamic control of redundantly actuated parallel kinematic manipulators.
Geometric trajectories if generated by means of trigonometric functions may show
inherent discontinuities regarding velocities and/or accelerations which may cause
problems for the drives. In order to overcome this issue and to generate C^2 continuous
refernce trajectories, we propose a novel technique consisting of modifying the motion
profile while preserving the overall geometric shape of the trajectory in the operational
space. Regarding the control strategy and to deal with the actuation redundancy, an
extended version of the PD controller with computed feedforward is proposed. The
computed control inputs, before being applied to the actuators, are first projected
using a regularization matrix based on the manipulator's kinematics in order to
remove the antagonistic internal forces. The overall proposed strategy including
the trajectory generator as well as the controller are experimentally validated on
the Dual-V robot, a three-degree-of-freedom redundantly actuated parallel kinematic
manipulator developed in our laboratory.

Keywords: Parallel manipulators, trajectory generation, dynamics, control.

1 Introduction

Parallel Kinematic Manipulators (PKMs) are mostly known for their superior dynamic
performance compared to their serial counterparts [1]. Indeed, in contrast with
serial manipulators in which the actuators are located on the moving links, the
actuators in the case of PKMs are located on the fixed base resulting in a much
lighter moving parts. Consequently, PKMs can achieve extremely high velocities and
accelerations [2]. Furthermore, the closed kinematic chains structure of PKMs yields
more stiffness, better accuracy and a higher load/weight ratio [1]. Nevertheless, PKMs
exhibit some drawbacks that may moderate their expansion in industrial applications.
The abundance of singularities and the relatively small workspace are the most
noteworthy limitations. While the latter is a matter of mechanical design, the former
can be solved through actuation redundancy. For this reason, Redundantly Actuated

M. Bennehar, A. Chemori, F. Pierrot and S. Krut: LIRMM, UMR 5506,CNRS Université de
Montpellier II, 161 rue Ada, 34392 Montpellier cedex 5, France, emails: ahmed.chemori@lirmm.fr,
mbennehar@gmail.com

De Gruyter Oldenbourg, ASSD – Advances in Systems, Signals and Devices, Volume 5, 2018, pp. 17–34.
DOI 10.1515/9783110470468-002

PKMs (RA-PKMs) have increasingly attracted the interest of researchers over the last two decades. Actuation redundancy significantly improves the performance of PKMs by homogenizing the dynamic properties throughout the workspace, allowing the manipulator to achieve very high accelerations in all configurations. However, despite the aforementioned qualities, PKMs potential is not yet completely explored and is still an open research area. From one hand, mechanical design, identification and optimization can be further investigated to achieve more efficient prototypes for industrial community. From the other hand, the constraints involved by the closed kinematic chains and the high nonlinear dynamics give rise to more challenging problems in terms of trajectory generation and control that earn to be studied.

Thanks to their extremely high dynamic capabilities, PKMs are typically used for high-speed industrial applications such as pick-and-place in food industry [3] and the assembly of electronic components [4]. Consequently, the most used solutions in terms of trajectory generation are based on traditional Point-to-Point (PtP) trajectories. In [5], an online smooth jerk-bounded trajectory generator using fifth order polynomials was proposed. The jerk boundedness yields an improved path tracking and a reduced wear on the robot. The problem of time optimality of PtP trajectories was addressed in [6]. An algorithm to derive the optimal trajectory was proposed and experimentally implemented on a 2-degree-of-freedom PKM. In [7], a variety of pick-and-place trajectory planners were evaluated with the aim of reducing vibrations of an elastic five-bar mechanism. Though the trend in this research area is towards PtP and pick-and-place trajectories, the inherent superior qualities of PKMs award them to be used in more complex modern industrial tasks such as laser cutting [8], machining [9] or even medical applications [10]. In this case, Geometric Closed Shape (GCS) trajectories draw more attention than traditional PtP ones [11]. This class of trajectories however, has not been sufficiently investigated in the literature. Indeed, in order to be better tracked by the robot's end effector (a traveling plate in the case of PKMs), the generated trajectories have to satisfy continuity constraint and respect the dynamic capabilities of the manipulator.

In order to track the reference trajectories with the best tracking performance, an efficient control scheme should take into account the nonlinear dynamics of the robot manipulator. Even if PKMs have different kinematics than serial manipulators, they actually share many dynamic similarities. Consequently, most of the developed control schemes for serial manipulators have been straightforwardly implemented on fully actuated PKMs [12–14]. However, in the case of RA-PKMs, a particular property characterizing this class of manipulators needs a special attention. Indeed, the non-uniqueness of the inverse dynamics solution yields antagonistic control forces that have no effect on the motion of the mechanical structure [15]. These control forces produce undesired internal pre-stress in the kinematic chains which may damage the manipulator, cause a loss of energy and generate mechanical vibrations. Consequently, a good control scheme should take into consideration such a phenomenon.

In this work, we address the two aforementioned issues namely, the continuous GCS trajectories generation and the internal forces free control of RA-PKMs. We propose herein a trajectory generator that takes into consideration continuity constraints in velocities and accelerations, both in task and joint spaces. Regarding the control solution, a joint space PD controller with computed feedforward [16] is proposed. This control scheme has the advantage of being low computationally efficient while compensating for some of the nonlinearities in the dynamics. In order to deal with the internal forces issue, the proposed controller is enhanced by a projection of the generated control torques with the seek of regularization. The proposed control scheme is then validated through real-time experiments on a 3-dof RA-PKM named Dual-V.

The rest of the paper is organized as follows. In section 2, the dynamic model of the Dual-V robot is presented. Section 3 is devoted to the proposed trajectory generator. The proposed control solution is detailed in Section 4. Experimental results are presented in Section 5. Finally, conclusions and perspectives are addressed in Section 6.

2 Description and Dynamic Modeling of the Dual-V Manipulator

Dual-V robot is a 3-dof planar RA-PKM belonging to the 4-RRR family. The arrangement of its four closed kinematic chains allows three independent movements for its traveling plate: two translations throughout the plane and one rotation about the z-axis. Hence, the Cartesian coordinates of the traveling plate can be described by the vector $X = [x, y, \theta]^T \in \mathbb{R}^3$. Regarding the dynamic modeling of the robot, the approach developed in [2] has been extended to take into account the rotational inertia of the forearms.

The input torque vector required to move the robot's mechanical structure $\Gamma \in \mathbb{R}^4$ is decomposed into three main sub-torques namely: γ_1, γ_2, $\gamma_3 \in \mathbb{R}^4$. Each sub-torque is responsible for moving specific parts of the mechanical structure of the robot as follows: the torque γ_1 is the required torque to move the traveling plate and a part of the couplers, it can be calculated using the equations of power of the actuators, it is given by:

$$\gamma_1 = J_m^{T*} M_I \ddot{X} \tag{1}$$

where $M_I \in \mathbb{R}^{3\times3}$ is the mass matrix of the traveling plate and a part of the cranks being expressed in Cartesian space, $J_m^{T*}(q) \in \mathbb{R}^{4\times3}$ is the pseudo-inverse of the transpose of the *inverse Jacobian matrix* J_m and $\ddot{X} \in \mathbb{R}^3$ is the Cartesian acceleration vector of the traveling plate. The sub-torque γ_2 is the required torque to move the cranks, the counter-masses and the remaining part of the couplers, it can be expressed as:

$$\gamma_2 = M_{II}\ddot{q} = M_{II}(\dot{J}_m\dot{X} + J_m\ddot{X}) \tag{2}$$

where $M_{II} \in \mathbb{R}^{4\times4}$ is the mass matrix including the dynamic parameters of the involved moving parts of the mechanical structure being expressed in joint space, $\dot{X} \in \mathbb{R}^4$ is the traveling plate velocity vector and $\dot{J}_m(q,\dot{q})$ is the time derivative of the inverse Jacobian matrix.

Up to now, only the inertia of the equivalent two-mass model [17] of each coupler is considered. In fact, the mass of each coupler is split up into two point-wise masses located at both ends of the considered link. Then, the rotational inertia of the two-mass model is considered instead of the real inertia of the couplers. Though this is a righteous assumption when light materials are used, it fails when the links are made with relatively heavy materials. This is the case of Dual-V parallel manipulator robot and hence, an additional term should be added. The sub-torque γ_3 is the additional torque component which accounts for the difference between real inertia and the one of the equivalent mass model of the couplers. Due to limitation on the number of pages, the details of this additional term are omitted and the interested reader is referred to [17] for a full description of the Dual-V robot dynamic modeling. It is worth noting though that $\gamma_3 = f(X, \dot{X}, \ddot{X})$ is a highly nonlinear function with respect to its arguments.

Finally, the full dynamic model of the Dual-V is obtained by summing the three sub-torques as follows:

$$\Gamma = J_m^{T*} M_I \ddot{X} + M_{II}(\dot{J}_m \dot{X} + J_m \ddot{X}) + \gamma_3 \tag{3}$$

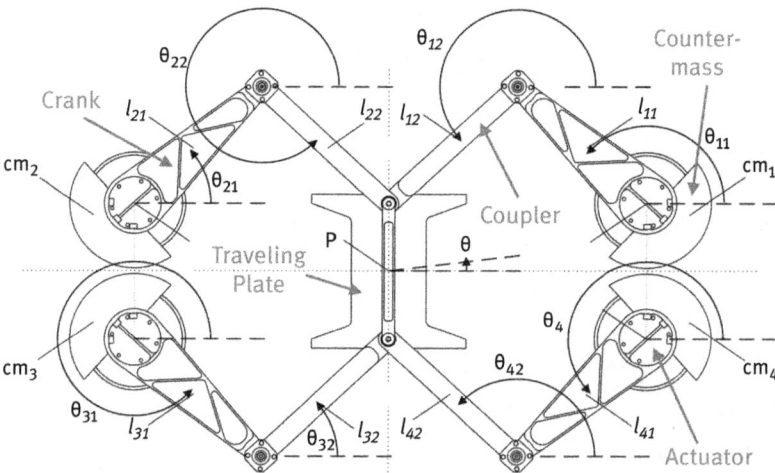

Fig. 1. CAD view of Dual-V parallel manipulator and its parameters definition.

Tab. 1. Dual-V geometric and dynamic parameters.

	Length [m]	Mass [Kg]	Inertia [Kgm²]
Crank	0.2800	1.169	0.012967
Coupler	0.2800	0.606	0.006417
Traveling plate	0.22	0.899	0.008168

The CAD model of the Dual-V with its parameters' definition is illustrated in Fig. 1; its geometric and dynamic parameters are summarized in Tab. 1. Further details regarding the Dual-V parallel manipulator can be found in [18] and [19].

3 C^2 Reference Trajectories Generation

Geometric Closed Shape trajectories are more complex than traditional PtP ones [5, 6] (e.g pick-and-place trajectories) and, in most cases, are generated by means of trigonometric functions parametrized by the time variable $t \in \mathbb{R}^+$. Generating trajectories with C^2 continuity constraints (i.e. on velocities and accelerations) may not be possible using standard analytical equations. Continuity of trajectories is a crucial constraint for the robot drives and actuators and is required in most robotic applications. Indeed, discontinuities in the trajectories may generate discontinuous control inputs and consequently, lead to undesired behavior of the mechanical structure such as mechanical vibrations, poor tracking performance and instabilities.

In this paper we are interested in the class of trajectories described by a sum of weighted sine and cosine functions. This class of trajectories has been chosen because it covers a large amount of geometric shapes from basic to the most complex ones (e.g. circles, ellipses, deltoids, . . .).

The general analytical form of the trajectories in question is then expressed by:

$$x(t) = \sum_{i=1}^{n} a_i \cos^{\alpha_i}\left(2\pi\frac{n_{1i}}{T}t\right) + b_i \sin^{\beta_i}\left(2\pi\frac{n_{2i}}{T}t\right) \tag{4}$$

where $T \in \mathbb{R}^+$ is the trajectory duration and a_i, α_i, b_i, β_i, $n_{ji} \in \mathbb{R}; i = 1, \ldots, n; j = 1, 2$ are scalars defining the overall shape of the trajectory.

In the sequel, we will first give an illustrative example of direct application of (4). Then, its major drawback is highlighted and a solution to overcome this issue is proposed.

3.1 Trajectories using Standard Geometric Functions

Let $x_d(t)$ be a reference Cartesian position defined by (4) to be tracked by the traveling plate of the robot. It is assumed throughout this brief that the reference trajectories are inside the workspace of the robot and hence are away from singularities. The reference velocity $v_d(t)$ and acceleration $a_d(t)$ for this motion are obtained by differentiating $x_d(t)$ with respect to time, giving thus:

$$v_d(t) = -\frac{2\pi}{T} \sum_{i=1}^{n} a_i \alpha_i n_{1i} \sin^{\alpha_i-1}\left(2\pi\frac{n_{1i}}{T}t\right) - b_i \beta_i n_{2i} \cos^{\beta_i-1}\left(2\pi\frac{n_{2i}}{T}t\right) \qquad (5)$$

$$a_d(t) = -\left(\frac{2\pi}{T}\right)^2 \sum_{i=1}^{n} a_i \alpha_i (\alpha_i-1) n_{1i}^2 \cos^{\alpha_i-2}\left(2\pi\frac{n_{1i}}{T}t\right)$$

$$+ b_i \beta_i (\beta_i-1) n_{2i}^2 \sin^{\beta_i-2}\left(2\pi\frac{n_{2i}}{T}t\right) \qquad (6)$$

Without loss of generality, assume that the manipulator starts and finishes its movement with zero velocity and acceleration (i.e. $v(0) = v(T) = 0$ and $a(0) = a(T) = 0$). However, using the original analytical functions (5) and (6) may lead to non-zero initial velocity and/or acceleration and thus, discontinuities in the generated motion. Indeed, if we replace $t = 0$ or $t = T$ in (5) and (6) we will obtain values dependent on initial and final conditions and thus, do not necessarily equal to zero. For instance, the obtained initial conditions for (5) and (6) are given by:

$$v_d(0) = -\frac{2\pi}{T} \sum_{i=1}^{n} b_i \beta_i n_{2i} = f_1(b_i, \beta_i, n_{2i}) \qquad (7)$$

and

$$a_d(0) = -\left(\frac{2\pi}{T}\right)^2 \sum_{i=1}^{n} a_i \alpha_i (\alpha_i-1) n_{1i}^2 = f_2(a_i, \alpha_i, n_{1i}) \qquad (8)$$

therefore, the resulting trajectories can be discontinuous in velocity and/or acceleration leading to discontinuous control torques and therefore big tracking errors and even problem of vibrations.

To further illustrate this inconvenience let's consider a simple circular trajectory with radius $r \in \mathbb{R}^+$ and a center whose coordinates are $(x_c, y_c) = (0, 0)$. The resulting trajectory needs to be C^2 continuous (in velocities and accelerations) both in operational and joint spaces in order to be appropriately tracked by the proposed controller. The analytical equations of positions, velocities and accelerations are given by:

$$\begin{cases} x_d(t) &= r\cos\left(\frac{2\pi}{T}(t-t_0)\right) \\ y_d(t) &= r\sin\left(\frac{2\pi}{T}(t-t_0)\right) \end{cases} \qquad (9)$$

$$\begin{cases} \dot{x}_d(t) &= -r\frac{2\pi}{T}\sin\left(\frac{2\pi}{T}(t-t_0)\right) \\ \dot{y}_d(t) &= r\frac{2\pi}{T}\cos\left(\frac{2\pi}{T}(t-t_0)\right) \end{cases} \qquad (10)$$

$$\begin{cases} \ddot{x}_d(t) & = & -r(\frac{2\pi}{T})^2 \cos\left(\frac{2\pi}{T}(t-t_0)\right) \\ \ddot{y}_d(t) & = & -r(\frac{2\pi}{T})^2 \sin\left(\frac{2\pi}{T}(t-t_0)\right) \end{cases} \tag{11}$$

These trajectories satisfy equations (4), (5) and (6) being only shifted in time by t_0. Figure 2 illustrates the plots of the obtained trajectories using (9), (10) and (11) for the parameters $t_0 = 0.5s$, $T = 1s$. It can be clearly noticed the presence of discontinuities on both velocity and acceleration profiles that have to be removed. In what follows, we propose a novel method to overcome this drawback and generate C^2 continuous reference trajectories.

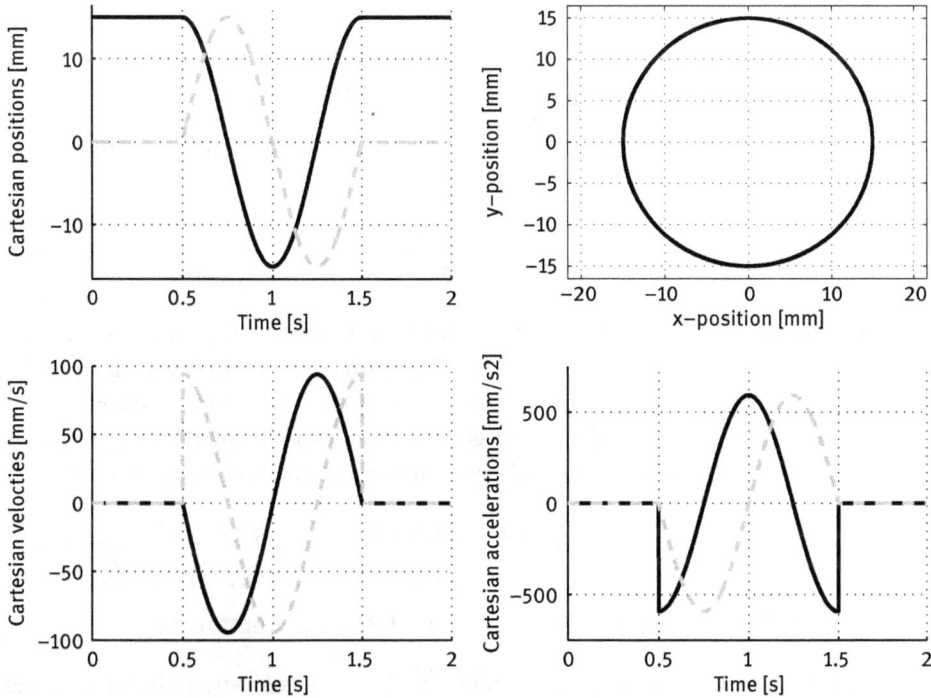

Fig. 2. View of the generated Cartesian trajectories using standard geometric functions. x_d (solid) and y_d (dashed).

3.2 Proposed C^2 Continuous Reference Trajectories

In order to overcome the discontinuity problem in the velocity and acceleration trajectories, we propose to revisit the analytical form of the desired operational trajectories. Consider again the previous illustrative example of the circular trajectory.

If we check the previous equations of the trajectory, it can be seen that they can be rewritten as follows:

$$\begin{cases} x_d(t) &= r\cos(\lambda(t)) \\ y_d(t) &= r\sin(\lambda(t)) \end{cases} \tag{12}$$

with $\lambda(t) = \frac{2\pi}{T}(t-t_0)$ which is a simple affine function of time t and when implemented, does not satisfy any continuity constraints on the velocities and accelerations. The corresponding velocities and accelerations can be obtained by differentiating (12) with respect to time, which gives:

$$\begin{cases} \dot{x}_d(t) &= -r\,\dot{\lambda}(t)\,\sin(\lambda(t)) \\ \dot{y}_d(t) &= r\,\dot{\lambda}(t)\,\cos(\lambda(t)) \end{cases} \tag{13}$$

$$\begin{cases} \ddot{x}_d(t) &= -r\left[\ddot{\lambda}(t)\sin(\lambda(t)) + \dot{\lambda}^2(t)\cos(\lambda(t))\right] \\ \ddot{y}_d(t) &= r\left[\ddot{\lambda}(t)\cos(\lambda(t)) - \dot{\lambda}^2(t)\sin(\lambda(t))\right] \end{cases} \tag{14}$$

From (13) and (14) one can notice that if $\lambda(t)$ is chosen such that it satisfies the following boundary constraints:

$$\begin{cases} \lambda(t_0) = 0, & \dot{\lambda}(t_0) = 0, & \ddot{\lambda}(t_0) = 0 \\ \lambda(t_0 + T) = 2\pi, & \dot{\lambda}(t_0 + T) = 0, & \ddot{\lambda}(t_0 + T) = 0 \end{cases} \tag{15}$$

one can ensure the continuity on the velocity and acceleration trajectories. One possible solution would then be to choose $\lambda(t)$ with a trapezoidal velocity profile. Indeed, this choice allows to specify boundary conditions and hence, the constraints (15) can be satisfied. A comparison between trapezoidal and linear velocity profile for $\lambda(t)$ is depicted in Fig. 3. The analytical equations of $\lambda(t)$ are now expressed as:

$$\begin{cases} \lambda(t) &= \frac{\lambda_f - \lambda_0}{2\tau^3(T-\tau)}[-t^4 + 2\tau t^3] + \lambda_0, & t_0 \le t < \tau \\ \lambda(t) &= \frac{\lambda_f - \lambda_0}{(T-\tau)}[t - \frac{\tau}{2}] + \lambda_0, & \tau \le t < T - \tau \\ \lambda(t) &= \frac{\lambda_f - \lambda_0}{2\tau^3(T-\tau)}[(t-T)^4 + 2\tau(t-T)^3] + \lambda_f, & \text{otherwise} \end{cases} \tag{16}$$

where τ is the duration of initial acceleration and final deceleration phases that has to be chosen according to the robot dynamic capabilities (choosing small values for τ may lead to unachievable high accelerations). λ_0 and λ_f are the initial and final values respectively for $\lambda(t)$ which depend on the type of the geometric trajectory. For instance, in the case of the previous circular trajectory example, $\lambda_0 = 0$ and $\lambda_f = 2\pi$.

The obtained trajectories using the proposed technique as depicted in Fig. 4 are now continuous in velocity and acceleration and result in the same circular trajectory in operational space as the original one. Therefore, they are more safe in the control scheme for the actuators of the robot manipulator.

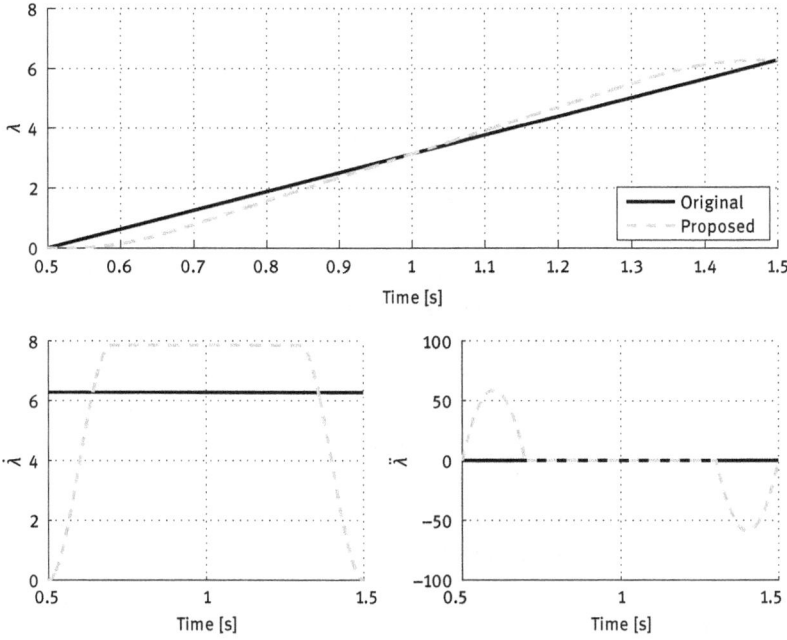

Fig. 3. Evolution of $\lambda(t)$, $\dot\lambda(t)$ and $\ddot\lambda(t)$ versus time.

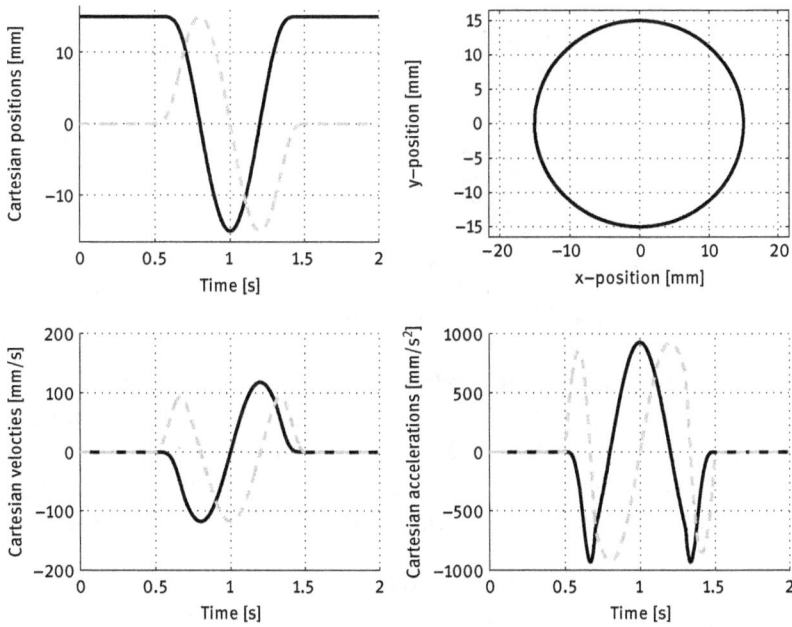

Fig. 4. Obtained C^2 continuous operational trajectories. x_d (solid) and y_d (dashed).

4 Proposed Control Solution

It is conventional that PKMs share many similarities with their serial counterparts regarding dynamic properties [14]. Consequently, the vast wide control literature developed for serial manipulators can be successfully applied to PKMs. However, in the particular case of RA-PKMs, the use of decentralized single axis controllers leads to antagonistic control efforts. Indeed, these efforts do not produce any motion of the manipulator. Thus, using conventional single axis strategies in the control loop will certainly involve internal forces (incompatible with the robot's kinematics) [15] creating internal pre-stress in the mechanism. The antagonistic forces can cause a multitude of undesirable phenomena such as loss of energy, instability and even mechanical vibrations. Consequently, the control architecture when designed has to take this issue in consideration. In this work we propose to enhance the well known PD controller with computed forward by projecting the computed inputs in order to reduce the effect of antagonistic forces. Figure 5 shows an overview of the different components of the proposed control scheme.

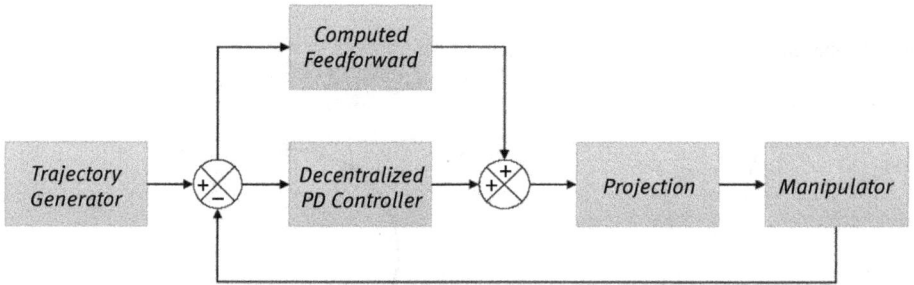

Fig. 5. Overview of the proposed control architecture.

4.1 Decentralized Joint Space PD Controller

Consider a RA-PKM with m degrees of freedom and n actuators and let the joints position vector be denoted by $q \in \mathbb{R}^n$. The PD control in joint space is a decentralized uncoordinated control strategy relying on the measured errors between the desired and the actual positions. Let $e(t) = q_d(t) - q(t)$ be the vector of joints errors, where $q_d(t)$ denotes the desired joint trajectory. Then, the PD control action is expressed by:

$$\Gamma_{PD} = K_p e(t) + K_d \dot{e}(t) \tag{17}$$

where K_p and K_d are positive feedback gains usually chosen as diagonal matrices. If all the actuators are identical, which is usually the case for PKMs, then the same feedback gains could be used for all axes, namely; $K_p = k_p I_{n \times n}$ and $K_d = k_d I_{n \times n}$, where $I_{n \times n}$ denotes the identity matrix and k_p, $k_d \in \mathbb{R}$ are positive scalars denoting the feedback gains that should be carefully tuned to achieve satisfactory tracking performance. A good accuracy is usually required in parallel robots; however, a PD controller is not able to assure that. Consequently, a computed feedforward is proposed to improve the tracking performance of the PD controller.

4.2 Tracking Accuracy Improvement through a Computed Feedforward

If the inverse dynamic model of the robot is known and its parameters are accurate enough, it would be interesting to exploit this knowledge in the control scheme to further improve the tracking performance. Indeed, the tracking errors can be significantly reduced by compensating the inherent nonlinear dynamics of the robot. One possible way would be the addition of a feedforward term. This strategy enables partial compensation of the nonlinear dynamics, i.e. only desired values of positions, velocities and accelerations are fed to the model-based term of controller. For the case of the Dual-V robot, the inverse dynamics are given by (3), hence, the feedforward control term can be expressed as:

$$\Gamma_{ff} = J_m^{T*}(q_d) M_I \ddot{X}_d + M_{II}(\dot{J}_m(q_d, \dot{q}_d) \dot{X}_d + J_m(q_d) \ddot{X}_d) + \gamma_{3_d} \tag{18}$$

where the subscript d refers to the desired quantities and $\gamma_{3_d} = f(X_d, \dot{X}_d, \ddot{X}_d)$.

4.3 Projection Method to Reduce Internal Forces

For RA-PKMs, the control inputs resulting from single axis controllers (i.e. PD portion) may contain antagonistic forces. These control inputs do not create any motion as they are in the null-space of J_m^T. However, they create uncontrolled internal pre-stress that may damage the manipulator. This issue has to be considered in the control architecture. One way to reduce the internal forces is by projecting the computed inputs onto the range of J_m^T by the following projector: [15]

$$R_{J_m^T} = I - N_{J_m^T} \tag{19}$$

where $N_{J_m^T} = I - J_m^{T*} J_m^T$ is the projector to the null-space of J_m^T. Hence, the internal antagonistic forces are eliminated by projecting the control torques onto the range space of J_m^T. This can be achieved using the projection matrix $R_{J_m^T}$ as:

$$\Gamma^* = R_{J_m^T}(\Gamma_{PD} + \Gamma_{ff}) \tag{20}$$

The overall block diagram of the proposed control scheme is illustrated in Fig. 6.

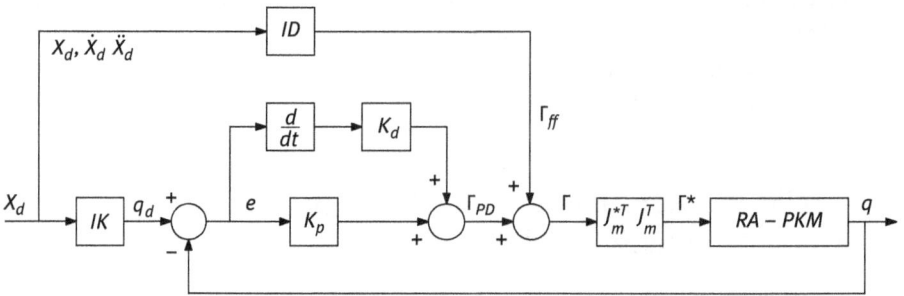

Fig. 6. Block diagram of the proposed control scheme; IK denotes the inverse kinematic model and ID the inverse dynamic model.

5 Real-Time Experimental Validation

5.1 Experimental Setup

All the links of the Dual-V are made of Aluminum. The cranks are mounted on four direct drive actuators manufactured by ETEL Motion Technology. The actuators are fixed on an Aluminum base and they can supply torques up to 127 N.m. Matlab software and Real-Time Workshop (both from Mathworks Inc.) have been used to implement and execute in real-time the proposed control scheme. The generated C code is then uploaded to the target PC (an industrial computer cadenced at 10 kHz and running xPC Target in real-time). The experimental setup is displayed in Fig. 7.

5.2 Obtained Experimental Results

The proposed trajectory generator and controller presented in the previous section were implemented on the Dual-V experimental testbed. For comparison purpose, we also implement the classical trajectories generator. In order to further investigate the benefits of the proposed approach, four different geometric shapes are included in the reference trajectories generator as illustrated in Fig. 8. The resulting reference trajectories include a circle, an ellipse, a tear-like and a deltoid geometric shapes. The traveling plate of the manipulator has to track each geometric trajectory in $T = 0.5\ s$. Between the end point of one closed-shape and the starting point of the next one,

Fig. 7. View of the experimental setup of Dual-V PKM; (1) the mechanical structure of the robot, (2) the host PC, (3) the target PC, (4) emergency stop, (5) host monitor, (6) xPC target monitor.

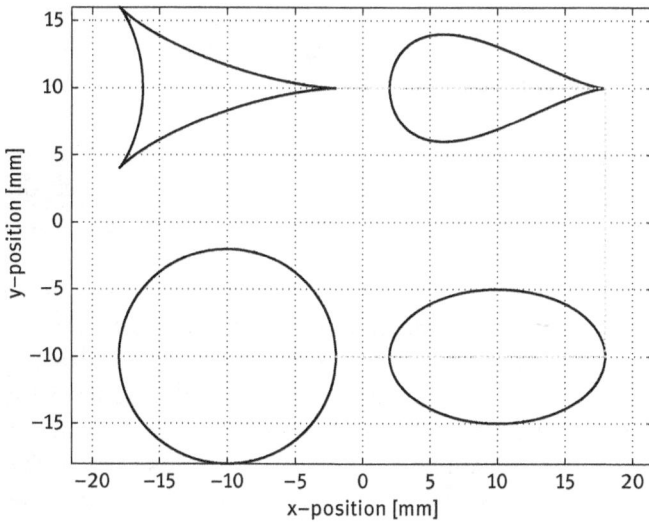

Fig. 8. Reference Cartesian trajectory used in experiments.

the end-effector has to perform a pick-and-place PtP trajectory defined by a 5th order polynomial function.

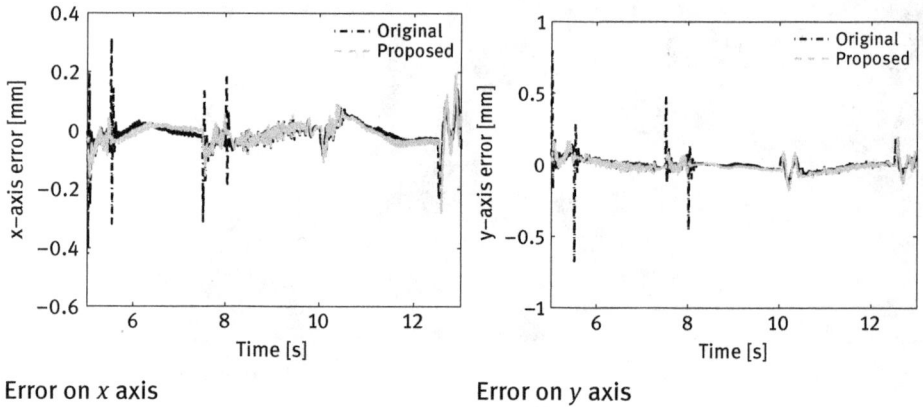

Error on x axis   Error on y axis

Fig. 9. Reference Cartesian trajectory used in experiments.

The feedback gains of the controller have been experimentally tuned using the trial-and-error technique. Usually, manipulators are not equipped with velocity sensors, therefore, velocities have to be numerically estimated from positions measurements. In our case, a carefully designed lead-lag filter is used to estimate the velocities.

Figure 9 displays the Cartesian tracking performance of the traveling plate on both x and y axes. It can be clearly seen that in the case of original trajectories, the discontinuities involve a poor tracking performance. This is more noticeable at the beginning and the end of the circle and the ellipse closed shapes which may lead to high frequency vibrations. However, if we check the obtained performance with the proposed approach, we can clearly observe the improvements in terms of tracking errors at the end and starting points. The proposed trajectories are accurately tracked thanks to the removed discontinuities. The improvements are more observable on the starting and end points of the circle and ellipse trajectories ($t = 5$, 5.5, 7.5 and $8s$). These results were expected since the main purpose of this paper was to propose a technique to remove these discontinuities from the trajectories generated by means of classical analytical functions. These discontinuities are a major source of tracking errors.

The generated control inputs when using the proposed method as well as those generated by the classical one are shown in Figs. 11 and 10 respectively. As expected, the controller generates very high torques when the reference trajectories contain discontinuities, because the drives need to generate high torques in order step from zero to a certain velocity in a very short time, which is in practice very delicate.

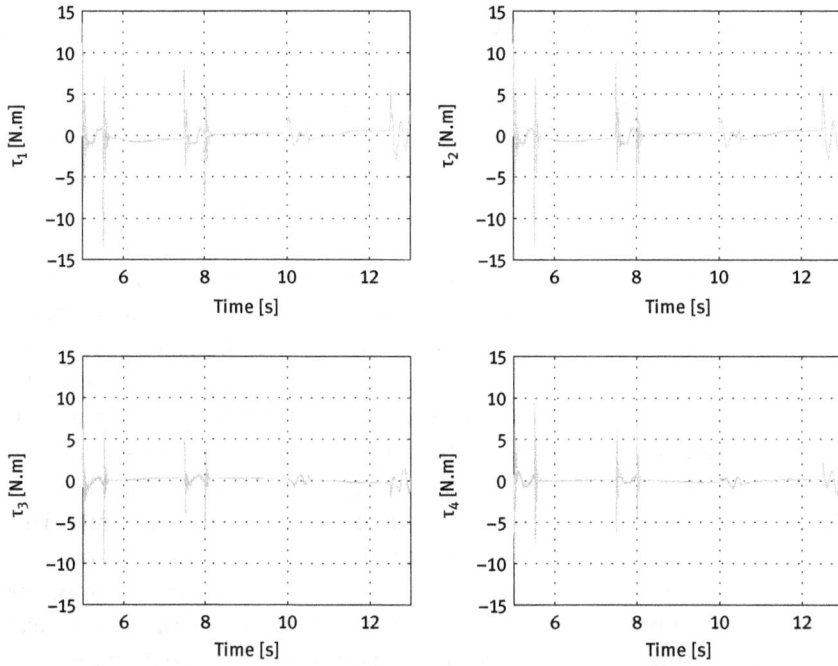

Fig. 10. Control inputs when using discontinuous trajectories.

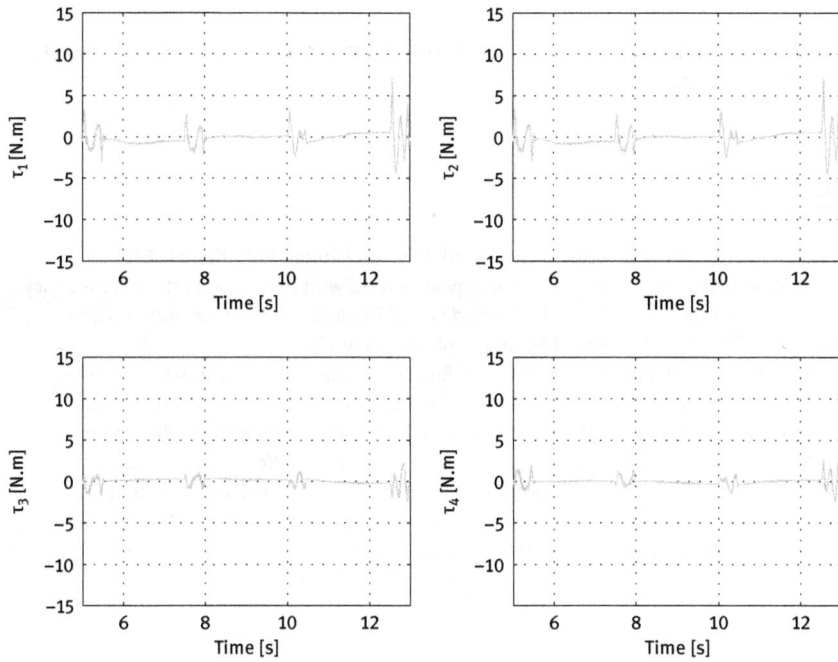

Fig. 11. Control inputs when using the proposed C^2 trajectories.

6 Conclusions and Future Work

Throughout this paper, we investigated the generation of continuous geometric closed-shape reference trajectories. Moreover, to accurately track these trajectories, a novel model-based controller in joint space has been proposed. The proposed controller enables the reduction of antagonistic internal forces caused by the unavoidable use of decentralized strategies in the control scheme. The geometric trajectories in question are defined using analytical functions that show inherent discontinuities regarding velocities and accelerations which may be a source of poor tracking. For this reason, we proposed in this work an approach to overcome this issue by modifying the motion profile without changing the overall geometric shape of the trajectory in operational space. Furthermore, to overcome the phenomenon of internal forces, a novel extended PD controller with computed feedforward was proposed. The computed control inputs were regularized using a projection matrix based on the manipulator's Jacobian matrix to reduce the antagonistic control inputs. This projection enables to significantly reduce the antagonistic internal forces that may damage the mechanical structure. This strategy allows to significantly reduce the energy consumption and to cancel the mechanical vibrations of the mechanical structure. To validate the proposed approaches, real-time experiments have been conducted on a 3-dof planar RA-PKM developed in our laboratory. The proposed trajectory generator demonstrated its superiority over the original one that utilizes discontinuous analytical equations.

Acknowledgments: This research was supported by the French National Research Agency, within the project ARROW.

Bibliography

[1] Jean-Pierre Merlet. *Parallel Robots*. Kluwer Academic Publishers, Netherlands, 2000.
[2] D. Corbel, M. Gouttefarde, and O. Company. Towards 100G with PKM. Is actuation redundancy a good solution for pick-and-place?. In *Proceedings of The IEEE Int. Conf. on Robotics and Automation (ICRA'10)*, 4675–4682, Anchorage-Alaska, May 2010.
[3] Ilian Bonev. Delta Parallel Robot – the Story of Success. *ParalleMIC*, May 2001.
[4] Marconi. The Gadfly manipulator. *Research Report 732*, Marconi Research Center, 1985.
[5] S. Macfarlane and E. A. Croft. Jerk-bounded manipulator trajectory planning: design for real-time applications. *IEEE Transactions on Robotics and Automation* 19(1):42–52, 2003.
[6] Y. Liu, C. Wang, J. Li and L. Sun. Time-Optimal Trajectory Generation of a Fast-Motion Planar Parallel Manipulator. In *Proceedings of The IEEE Int. Conf. on Intelligent Robots and Systems (IROS'06)*, 754–759, Beijing-China, October 2006.

[7] C. Barnard, S. Briot and S. Caro. Trajectory generation for high speed pick and place robots. In *Proceedings of The ASME 2012 11th Biennial Conference On Engineering Systems Design And Analysis*, Nantes-France, July 2012

[8] L. E. Bruzzone, R.M Molfino and R. P. Razzoli. Modeling and design of a parallel robot for laser-cutting applications. In *Proceedings of the IASTED International Conference on Modelling, Identification, and Control (MIC 2002)*, 518–522, Innsbruck-Austria, February 2002.

[9] H. Chanal. *Etude de l'emploi des machines outils à structure parallèle en usinage*, PhD thesis, Université Blaise Pascal, Clermont-Ferrand-France, 2006.

[10] Y. LI and Q. XU. Design and development of a medical parallel robot for cardiopulmonary resuscitation. *IEEE/ASME Transactions on Mechatronics*, 12(3):265–273, 2007.

[11] H. J. Su, P. Dietmaier and J. M. McCarthy. Trajectory planning for constrained parallel manipulators. *Journal of mechanical design*, 125(4):709–716, 2003.

[12] F. Paccot, N. Andreff and P. Martinet. A Review on the Dynamic Control of Parallel Kinematic Machines: Theory and Experiments. *The International Journal of Robotics Research*, 28(3):395–416, 2009.

[13] J. F. He, H. Z. Jiang, D. C. Cong, Z. M. Ye and J. W. Han. A Survey on Control of Parallel Manipulator. *Key Engineering Materials*, 339:307–313, 2007.

[14] H. Cheng, Y. K. Yiu and Z. Li. Dynamics and Control of Redundantly Actuated Parallel Manipulators. *IEEE/ASME Transactions on Mechatronics*, 8(4):483–491, 2003.

[15] A. Müller and T. Hufnagel. A projection method for the elimination of contradicting control forces in redundantly actuated PKM. In *Proceedings of The IEEE Int. Conf. on Robotics and Automation (ICRA'11)*, 3218–3223, Shanghai-China, May 2011.

[16] F. Reyes and R. Kelly. Experimental evaluation of model-based controllers on a direct-drive robot arm. *Mechatronics*, 11(3):267–282, 2001.

[17] V. Van Der Wijk, S. Krut, F. Pierrot and J. L. Herder. Design and experimental evaluation of a dynamically balanced redundant planar 4-RRR parallel manipulator. *The International Journal of Robotics Research*, 32(6):744–759, 2013.

[18] J. Wu, J. Wang and Z. You. A comparison study on the dynamics of planar 3-DOF 4-RRR, 3-RRR and 2-RRR parallel manipulators. *Robotics and Computer Integrated Manufacturing*, 27(1):150–156, 2011.

[19] V. Van Der Wijk, S. Krut, F. Pierrot and J. L. Herder. Generic Method for Deriving the General Shaking Force Balance Conditions of Parallel Manipulators with Application to a Redundant Planar 4-RRR Parallel Manipulator. In *Proceedings of The 13th World Congress in Mechanism and Machine Science*, Guanajuato-Mexico, 2011.

Biographies

Moussab Bennehar received his B.Sc. and M.Sc. degrees in electronics from the university of Constantine, Algeria, in 2012. He is currently a Robotics PhD student at Laboratoire d'Informatique, de Robotique et de Microélectronique de Montpellier (LIRMM - UMR 5506) in the Montpellier Laboratory of Computer Science, Robotics, and Microelectronics, France. His research is focused on trajectory generation and control of multibody systems. His PhD thesis is focused on advanced control techniques of parallel kinematic manipulators.

Ahmed Chemori received his M.Sc. and Ph.D. degrees, respectively in 2001 and 2005, both in automatic control from the Grenoble Institute of Technology. He has been a post-doctoral fellow with the automatic control laboratory of Grenoble in 2006. He is currently a tenured research scientist in Automation and Robotics at the Montpellier Laboratory of Computer Science, Robotics, and Microelectronics. His research interests include nonlinear, adaptive and predictive control and their applications in humanoid robotics, underactuated systems, parallel robots and underwater vehicles.

Sébastien Krut received the M.S. degree in mechanical engineering from the Pierre and Marie Curie University, Paris, France, in 2000 and the Ph.D. degree in automatic control from the Montpellier University of Sciences, Montpellier, France, in 2003. He has been a Post-doctoral fellow with the Joint Japanese-French Robotics Laboratory (JRL) in Tsukuba, Japan in 2004. He is currently a tenured research scientist in Robotics for the French National Centre for Scientific Research (CNRS), at the Montpellier Institute of Informatics, Microelectronics and Robotics (LIRMM in French), Montpellier, France. His research interests include design and control of robotic systems in general, and more specifically of parallel manipulators.

François Pierrot is a senior researcher in robotics for CNRS. His research interests include the creation of innovative robots and he considers both mechanical design and control strategies.

S. Ben Atia, A. Messaoud and R. Ben Abdennour

Decoupled Multimodel Predictive Control Based on Multi-observer for Discrete-time Uncertain Nonlinear Systems

Abstract: This paper deals with a decoupled multimodel predictive control for discrete-time uncertain nonlinear systems. The control scheme is based on a multi-observer for the state estimation of uncertain nonlinear systems described by decoupled multimodel. A partial controller and observer is synthesized for each local model. In order to ensure the closed-loop performances, a supervisor is proposed to select the appropriate controller. Simulation example is carried out to exhibit the effectiveness of the proposed control strategy.

Keywords: Uncertain nonlinear systems, Decoupled multimodel, Multi-observer, Predictive control, Supervisor.

1 Introduction

The strategy of predictive control has become a method increasingly used in many application areas. The basic idea is to use knowledge from the model to envisage various scenarios of the system operation in the future and choose the best according to the objectives.

The implementation of model predictive control requires the calculation of a sequence of future controls that are obtained by the minimisation of a quadratic performance criterion over a finite horizon.

For systems with nonlinear models, this control has several limits, namely the complexity of the model considered in the optimisation problem. Otherwise, the efficiency of the predictive control is closely related to the accuracy of the model that represents the system behavior. Indeed, the system modeling is based on physics' equations or identification techniques to obtain the so called nominal model. However, this model is often achieved at the cost of many simplifications involved in the system model, or on the influence of the outside environment. Thus, the nominal model does not accurately translate the dynamic behavior of the real system, hence the need to take into account modeling uncertainties.

S. Ben Atia, A. Messaoud and R. Ben Abdennour: Research Unit: Numerical Control of Industrial Processes (CONPRI), University of Gabes, National School of Engineers of Gabes (ENIG), Omar Ibn Khattab Street, 6029 Gabes, Tunisia, email: xxx@xxx

De Gruyter Oldenbourg, ASSD – Advances in Systems, Signals and Devices, Volume 5, 2018, pp. 35–54.
DOI 10.1515/9783110470468-003

The state estimation and the control of uncertain linear and nonlinear systems has been the subject of numerous papers [1, 3, 5].

In [4] and [21], the authors discussed the robustness of the model predictive control (MPC) for discrete-time uncertain systems with time-delay. A robust fuzzy model predictive control joining the MPC strategy and the Takagi–Sugeno (T–S) model was discussed in [9–11, 18, 19].

In fact, systems with nonlinear behaviors and complex structures are frequently encountered in practice. Indeed, the handling of these systems is a difficult task in many practical situations, especially when they have highly nonlinear dynamic behaviors. Consequently, it is difficult to synthesize control laws or to implement strategies for system diagnosis for these systems with mathematical complexity.

In the literature, the linearisation method is widely used [20]. This method consists in the linearisation of the nonlinear system around operating points. However, the global behavior of the system is not fully represented by the linearised model especially when the system has high nonlinearities. Subsequently, it is fundamental to ensure the compromise between a good characterisation of the system by a method that describes accurately the system behavior in the different operating ranges and a method that allows the exploitation of the techniques extensively developed in the linear case.

The multimodel approach represents an efficient alternative for the modeling of nonlinear systems. There are two main types of multimodel, namely the coupled (T–S model) and decoupled multimodel. The main difference between these two models resides in the way of local models' combination. Indeed, the coupled structure relies on fuzzy rules and it consists of partial models sharing the same state vector. On the other hand, the decoupled multimodel includes decoupled (independent) partial models, which means that each local model has its own state vector. In order to obtain a multimodel, three different methods can be distinguished; the linearisation method [20], the convex polytopic transformation [6] and identification [2, 13]. The last method is retained in this work for the obtention of a decoupled multimodel described in the state space.

The implementation of many control laws (such as state feedback control, model predictive control...) as well as the supervision and fault detection require usually the use of the state representation. However, the design of this strategies is often carried out assuming the availability of the state variables. In practice, this assumption is unrealistic. Hence, the design of an observer for the estimation of the unmeasurable states is an unavoidable task to design any control strategy that depends on the state representation.

Over the last decades, state estimation of nonlinear systems described by T-S have been widely investigated through several research works. However, the design of observers for decoupled multimodel was not extensively discussed in the literature. In fact, the state estimation of nonlinear systems represented by decoupled multimodel was recently treated in few works [12, 14].

In [16], the author designed a robust observer for continuous and discrete-time uncertain nonlinear systems described by decoupled multimodel.

In the present paper, a decoupled multimodel predictive control is investigated for the control of a large class of discrete-time uncertain nonlinear systems.This control scheme is based on a multi-observer for the estimation of state variables. The efficiency of the designed multi-controller is conditioned by the accuracy of the state estimation provided by the proposed multi-observer.

This paper is organized as follows. Section two is devoted to the design of a robust multi-observer for the state estimation of uncertain decoupled multimodel. The formulation of the partial model predictive control is presented in section three. Section four is reserved to the development of a supervised decoupled multimodel predictive control for discrete-time uncertain nonlinear systems. An illustrative example is carried out in section five to show the significance of the proposed control scheme. Section six concludes the paper.

2 Robust Multi-Observer Design for Uncertain Decoupled Multimodel

2.1 Uncertain Decoupled Multimodel

In this part, the discrete-time uncertain decoupled multimodel is presented. Two types of uncertainties can be distinguished; unstructured and structured uncertainties. In this work, the retained form of uncertainties is the structured norm-bounded uncertainties.

The multimodel representation of a discrete-time uncertain nonlinear system is described by the following uncertain decoupled multimodel:

$$
\begin{cases}
x_i(k+1) = (A_i + \Delta A_i(k)) x_i(k) + (B_i + \Delta B_i(k)) u(k) \\
y_{mi}(k) = C_i x_i(k) \\
y_m(k) = \sum_{i=1}^{N_m} \mu_i(z(k)) y_{mi}(k)
\end{cases}
\tag{1}
$$

where $x_i \in \mathbb{R}^{n_i}$ is the state vector and $y_{mi} \in \mathbb{R}^p$ is the output of the ith partial model.

$u \in \mathbb{R}^m$ and $y_m \in \mathbb{R}^p$ are the control signal and the output of the system.

The matrices $A_i \in \mathbb{R}^{n_i \times n_i}$, $B_i \in \mathbb{R}^{n_i \times m}$ and $C_i \in \mathbb{R}^{p \times n_i}$ are known constant matrices. N_m is the number of partial models.

The contribution of each partial model is defined by the weighting functions $\mu_i(z(k))$.

$$\mu_i(z(k)) = \frac{e^{(\frac{-(z(k)-c_i)^2}{\sigma_d^2})}}{\sum\limits_{j=1}^{N_m} e^{(\frac{-(z(k)-c_i)^2}{\sigma_d^2})}}, \quad i = 1, 2, \ldots, N_m$$

They have the following properties:

$$\sum_{i=1}^{N_m} \mu_i(z(k)) = 1, \quad 0 \le \mu_i(z(k)) \le 1, \quad \forall i = 1, \ldots, N_m, \quad \forall k$$

z is the decision variable. It depends on the measurable or unmeasurable states, the input signal or the system output. The centers and the dispersion are c_i and σ_d, respectively.

The parametric uncertainties are the unknown matrices that are assumed to be of the form:

$$\begin{aligned}\Delta A_i(k) &= D_{Ai}\Delta_{Ai}(k)E_{Ai} \\ \Delta B_i(k) &= D_{Bi}\Delta_{Bi}(k)E_{Bi}, \quad i = 1 \cdots N_m\end{aligned} \tag{2}$$

D_{Ai}, D_{Bi}, E_{Ai} and E_{Bi} are known constant matrices of appropriate dimensions.
$\Delta_{Ai}(k)$ and $\Delta_{Bi}(k)$ are unknown and time-varying matrix functions satisfying the following conditions:

$$\begin{aligned}\Delta_{Ai}^T(k)\Delta_{Ai}(k) &\le I \\ \Delta_{Bi}^T(k)\Delta_{Bi}(k) &\le I, \quad i = 1 \cdots N_m\end{aligned} \tag{3}$$

The compact form of the equation (1) can be obtained by defining the vector $x(k)$ as follows:

$$x(k) = \left[x_1^T(k) \cdots x_i^T(k) \cdots x_{N_m}^T(k)\right]^T \in \mathbb{R}^n, \quad n = \sum_{i=1}^{N_m} n_i \tag{4}$$

The equation (1) can be written in the following compact form:

$$\begin{cases} x(k+1) = (\mathbb{A} + \Delta\mathbb{A}(k))x(k) + (\mathbb{B} + \Delta\mathbb{B}(k))u(k) \\ y_m(k) = \mathbb{C}(k)x(k) \end{cases} \tag{5}$$

where the matrices $\mathbb{A} \in \mathbb{R}^{n \times n}$, $\mathbb{B} \in \mathbb{R}^{n \times m}$ et $\mathbb{C} \in \mathbb{R}^{p \times n}$ are defined as follows:

$$\mathbb{A} = diag\{A_1 \cdots A_{N_m}\}, \mathbb{B} = \begin{bmatrix} B_1 & \cdots & B_i & \cdots & B_{N_m} \end{bmatrix}$$
$$\mathbb{C}(k) = \begin{bmatrix} \mu_1(k)C_1^T & \cdots & \mu_i(k)C_i^T & \cdots & \mu_{N_m}(k)C_{N_m}^T \end{bmatrix}$$

The matrix $\mathbb{C}(k)$ can be written as follows:

$$\mathbb{C}(k) = \sum_{i=1}^{N_m} \mu_i(z(k))C_i$$

with $\mathbb{C}_i$ is a block matrix of the form:

$$\mathbb{C}_i = [0 \dots C_i \dots 0]$$

The uncertainties are given by:

$$\begin{aligned}
\Delta\mathbb{A}(k) &= \mathbb{D}_A \Delta_\mathbb{A}(k) \mathbb{E}_A \\
\Delta\mathbb{B}(k) &= \mathbb{D}_B \Delta_\mathbb{B}(k) \mathbb{E}_B
\end{aligned} \tag{6}$$

with $\mathbb{D}_j$ and $\mathbb{E}_j$ $(j = A, B)$ are augmented matrices defined as follows:

$$\mathbb{D}_A = diag\{D_{A1} \quad \cdots \quad D_{AN_m}\} \tag{7}$$
$$\mathbb{D}_B = diag\{D_{B1} \quad \cdots \quad D_{BN_m}\} \tag{8}$$
$$\mathbb{E}_A = diag\{E_{A1} \quad \cdots \quad E_{AN_m}\} \tag{9}$$
$$\mathbb{E}_B = \begin{bmatrix} E_{B1}^T & \cdots & E_{BN_m}^T \end{bmatrix} \tag{10}$$

The matrix functions are given by the following compact form:

$$\Delta_\mathbb{A}(k) = diag\{ \Delta_{A_1}(k) \quad \cdots \quad \Delta_{AN_m}(k) \} \tag{11}$$
$$\Delta_\mathbb{B}(k) = diag\{ \Delta_{B_1}(k) \quad \cdots \quad \Delta_{BN_m}(k) \} \tag{12}$$

2.2 State Estimation of Uncertain Decoupled Multimodel

The state estimation of the system (1) is ensured by the following multiobserver:

$$\begin{cases} \hat{x}(k+1) = \mathbb{A}\hat{x}(k) + \mathbb{B}u(k) + \mathbb{L}(y(k) - \hat{y}(k)) \\ \qquad\quad \hat{y}(k) = \mathbb{C}(k)\hat{x}(k) \end{cases} \tag{13}$$

with $\mathbb{L} = \begin{bmatrix} L_1 & \cdots & L_i & \cdots & L_{N_m} \end{bmatrix}^T \in \mathbb{R}^n$

The multiobserver synthesis problem can be formulated as adjusting the multi-observer gain such that the state estimation error given by equation (14) converges to zero. So, this gain should be adjusted to provide accurate state estimation in the presence of parametric uncertainties.

The dynamic of the estimation error is given as follows:

$$\begin{aligned}
\tilde{x}(k+1) &= x(k+1) - \hat{x}(k+1) \\
&= (\mathbb{A} - \mathbb{L}\mathbb{C}(k))\tilde{x}(k) + \Delta\mathbb{A}(k)x(k) + \Delta\mathbb{B}(k)u(k)
\end{aligned} \tag{14}$$

Assuming that:

$$\xi(k) = \begin{bmatrix} \tilde{x}(k) \\ x(k) \end{bmatrix} \tag{15}$$

Then, the dynamic of the augmented system has the following expression:

$$\xi(k+1) = \Gamma(k)\xi(k) + Y(k)u(k) \tag{16}$$

with

$$\Gamma(k) = \begin{bmatrix} \mathbb{A} - \mathbb{L}\mathbb{C}(k) & \Delta\mathbb{A}(k) \\ 0 & \mathbb{A} + \Delta\mathbb{A}(k) \end{bmatrix}$$

$$Y(k) = \begin{bmatrix} \Delta\mathbb{B}(k) \\ \mathbb{B} + \Delta\mathbb{B}(k) \end{bmatrix}$$

Consider the following Lyapunov functional:

$$V(k) = \xi^T(k)P\xi(k) \tag{17}$$

with $P = \begin{bmatrix} P_1 & 0 \\ 0 & P_2 \end{bmatrix}$.

The convergence of the estimation error is ensured the forward difference of Lyapunov functional is negative ($\Delta V(k) = V(k+1) - V(k) < 0$). The convergence conditions are formulated in terms of linear matrix inequalities and are given by the following theorem.

THEOREM [15] The estimation error between the considered multimodel (5) and the multi-observer (13) is asymptotically stable if there exist symmetric matrices P_1 et P_2, a matrix X of appropriate dimension and positive scalars ρ_1 and ρ_2 such that the following linear matrix inequalities hold:

$$\begin{bmatrix} -P_1 & 0 & 0 & \Omega_i^T & 0 & 0 & 0 \\ 0 & \rho_1 \mathbb{E}_A^T \mathbb{E}_A - P_2 & 0 & 0 & \mathbb{A}^T P_2 & 0 & 0 \\ 0 & 0 & \rho_2 \mathbb{E}_B^T \mathbb{E}_B & 0 & \mathbb{B}^T P_2 & 0 & 0 \\ \Omega_i & 0 & 0 & -P_1 & 0 & P_1 \mathbb{D}_A & P_1 \mathbb{D}_B \\ 0 & P_2 \mathbb{A} & P_2 \mathbb{B} & 0 & -P_2 & P_2 \mathbb{D}_A & P_2 \mathbb{D}_B \\ 0 & 0 & 0 & \mathbb{D}_A^T P_1 & \mathbb{D}_A^T P_2 & -\rho_1 I & 0 \\ 0 & 0 & 0 & \mathbb{D}_B^T P_1 & \mathbb{D}_B^T P_2 & 0 & -\rho_2 I \end{bmatrix} < 0 \tag{18}$$

with:

$$\Omega_i = P_1 \mathbb{A} - X\mathbb{C}_i \tag{19}$$

$\mathbb{L} = P_1^{-1}X$ is the multi-observer gain.

3 Partial Model Predictive Control Formulation

The philosophy of the model predictive control is to use the model to predict the system behavior and choose the best decision, in the sense of a certain cost. Thus, the control synthesis relies on the minimisation of a cost function J_i that leads to the optimal control:

$$J_i = \frac{1}{2} \left\{ \sum_{j=1}^{N_p} \|\hat{y}_i(k+j/k) - y_c(k+j)\|_v^2 + \sum_{j=0}^{N_u-1} \|\Delta u_i(k+j)\|_\eta^2 \right\} \tag{20}$$

with:

N_p and N_u are the prediction horizon and the control horizon, respectively.

v and η are positive weighting factors.

$y_c(k+j/k)$ is the j steps reference trajectory and $\hat{y}_i(k+j/k)$ is the j steps predictor output.

$\Delta u_i(k+j)$ are the control increments for j steps ahead with:

$\Delta u_i(k+j) = 0 \ \forall j \in [N_u, N_p]$.

Based on equation (1), the corresponding estimated state, for each partial model, predicted for j steps ahead can be written as follows:

$$\begin{cases} \hat{x}_i(k+j) = A_i \hat{x}_i(k+j-1) + B_i u_i(k+j-1) \\ \hat{y}_i(k+j) = C_i \hat{x}(k+j) \end{cases} \quad \forall j = 1, \ldots, N_p \tag{21}$$

The partial model output $\hat{y}_i(k+j)$ predicted for j steps ahead is given by the following expression written in a matrix form:

$$\hat{Y}_i(k) = \Xi_i \hat{x}_i(k) + \Psi_i U_i(k) \tag{22}$$

with

$$\Xi_i = \begin{bmatrix} C_i A_i \\ C_i A_i^2 \\ \vdots \\ C_i A_i^{N_p} \end{bmatrix}, \quad \Psi_i = \begin{bmatrix} C_i B_i & 0 & \cdots & 0 \\ C_i A_i B_i & \ddots & \ddots & \vdots \\ \vdots & \ddots & \ddots & 0 \\ C_i A_i^{N_p-1} B_i & \cdots & C_i A_i B_i & C_i B_i \end{bmatrix}$$

$$U_i(k) = \begin{bmatrix} u_i(k) & u_i(k+1) & \cdots & u_i(k+N_p-1) \end{bmatrix}^T$$

$$\hat{Y}_i(k) = \begin{bmatrix} \hat{y}_i(k+1) & \hat{y}_i(k+2) & \cdots & \hat{y}_i(k+N_p) \end{bmatrix}^T$$

The equation (20) written in a matrix form is given by the following expression:

$$\begin{aligned} J_i &= \frac{1}{2} \left[\|\hat{Y}_i(k) - Y_c(k)\|_\vartheta^2 + \|\Delta U_i(k)\|_\aleph^2 \right] \\ &= \frac{1}{2} \left(\left(\hat{Y}_i(k) - Y_c(k)\right)^T \vartheta \left(\hat{Y}_i(k) - Y_c(k)\right) + \Delta U_i^T(k) \aleph \Delta U_i(k) \right) \end{aligned} \tag{23}$$

with

$$Y_c(k) = \begin{bmatrix} y_c(k+1) & y_c(k+2) & \cdots & y_c(k+N_p) \end{bmatrix}^T$$

$$\Delta U_i(k) = \left[\begin{array}{ccccccc} \Delta u_i(k) & \cdots & \Delta u_i(k+N_u-1) & \overbrace{0 \quad \cdots \quad 0}^{N_p-N_u} \end{array} \right]^T$$

$$\vartheta = \left[\begin{array}{cccc} v & 0 & \cdots & 0 \\ 0 & \ddots & \ddots & \vdots \\ \vdots & \ddots & \ddots & 0 \\ 0 & \cdots & 0 & v \end{array} \right], \quad \aleph = \overbrace{\left[\begin{array}{cccccc|cccc} \multicolumn{6}{c}{\overbrace{}^{N_u}} \\ 2\eta & -\eta & 0 & \cdots & 0 & 0 & \cdots & 0 \\ -\eta & 2\eta & \ddots & \ddots & \vdots & \vdots & \cdots & \vdots \\ 0 & \ddots & \ddots & \ddots & 0 & \vdots & \cdots & \vdots \\ \vdots & \ddots & \ddots & 2\eta & -\eta & \vdots & \cdots & \vdots \\ 0 & \cdots & 0 & -\eta & \eta & 0 & \cdots & 0 \\ 0 & \cdots & \cdots & \cdots & 0 & 0 & \cdots & 0 \\ \vdots & \ddots & \ddots & \ddots & \ddots & \ddots & \ddots & \vdots \\ 0 & \cdots & \cdots & \cdots & \cdots & 0 & \cdots & 0 \end{array} \right]}^{N_p}$$

The partial criterion J_i has the following expression:

$$J_i = \frac{1}{2} \left(U_i^T(k) \Psi_i^T \vartheta \Psi_i U_i(k) + 2U_i^T(k) \left(\Psi_i^T \vartheta \Xi_i \hat{x}_i(k) - \Psi_i^T \vartheta Y_c(k) \right) \right)$$

$$+ \frac{1}{2} \left(U_i^T(k) \aleph U_i(k) - 2\eta u_i(k) u_i(k-1) + ct_i \right)$$

$$= \frac{1}{2} U_i^T(k) \Theta_i U_i(k) + U_i^T(k) \varphi_i + ct_i \tag{24}$$

where

$$\Theta_i = \Psi_i^T \vartheta \Psi_i + \aleph$$

$$\varphi_i = \left[\begin{array}{cc} \Psi_i^T \vartheta \Xi_i & -\Psi_i^T \vartheta \end{array} \right] \left[\begin{array}{c} \hat{x}_i(k) \\ Y_c(k) \end{array} \right] - \left[\begin{array}{c} \eta u_i(k-1) \\ 0 \\ \vdots \\ 0 \end{array} \right]$$

$ct_i = (\Xi_i \hat{x}_i(k) - Y_c(k))^T \vartheta (\Xi_i \hat{x}_i(k) - Y_c(k)) + \eta u_i^2(k-1)$ is a constant independent from $U_i(k)$.

$\dfrac{\partial J_i}{\partial U_i(k)}$ is the partial derivative of the criterion J_i, its expression is given as follows:

$$\frac{\partial J_i}{\partial U_i(k)} = \Theta_i U_i(k) + \varphi_i \tag{25}$$

The optimal control law has the following expression:

$$U_i(k) = -\Theta_i^{-1}\varphi_i \tag{26}$$

4 Decoupled Multimodel Predictive Control with Supervisor

The efficiency of the model predictive control is closely related to the accuracy of the system model. In the case of unmeasurable states, the accuracy of the model depends on the effectiveness of the multi-observer synthesized for the state estimation. The present work presents a multi-observer that serves to reconstruct the state variables of discrete-time uncertain decoupled multimodel. This multi-observer is used in the elaboration of the control law.

The control scheme contains three blocks (Fig. 1). The multi-observer/ multi-controller block is designed to estimate the state variables that serves to elaborate the local control laws of the multi-controller satisfying the closed-loop objectives. The supervision block include a set of partial predictors that describe the system behavior in each operating zone, it has as role to select the convenient partial predictor and consequently the corresponding partial controller. The adequate local controller is chosen depending on a performance criterion $Jc_i(k)$ that is based on the error between the system and the partial predictors' outputs. The switching block selects the suitable local controller corresponding to the minimal criterion value. The selection of the suitable controller requires the minimisation of the criterion $Jc_i(k)$ given by the following expression [7, 8, 17]:

$$Jc_i(k) = \alpha\varepsilon_i^2(k) + \beta \sum_{j=1}^{k} e^{-\lambda(k-j)}\varepsilon_i^2(j), \quad i = 1\ldots N_m \tag{27}$$

with

$\varepsilon_i(k)$ $(\varepsilon_i(k) = y(k) - \hat{y}_{ci}(k))$ denotes the error between the real and the ith predictor output.

α and β are positive parameters denoting the weighting factors of the instantaneous and the long-term measures accuracy.

λ is the forgetting factor.

N_m is the number of the local controllers.

The ith predictor output is obtained by the inversion of the ith partial controller and it given as follows:

$$\hat{Y}_{ci}(k) = \left(\Psi_i^T\vartheta\right)^{-1}\left(\Psi_i^T\vartheta\Psi_i + \aleph\right)U_i(k) + \Xi_i\hat{x}_i(k)$$

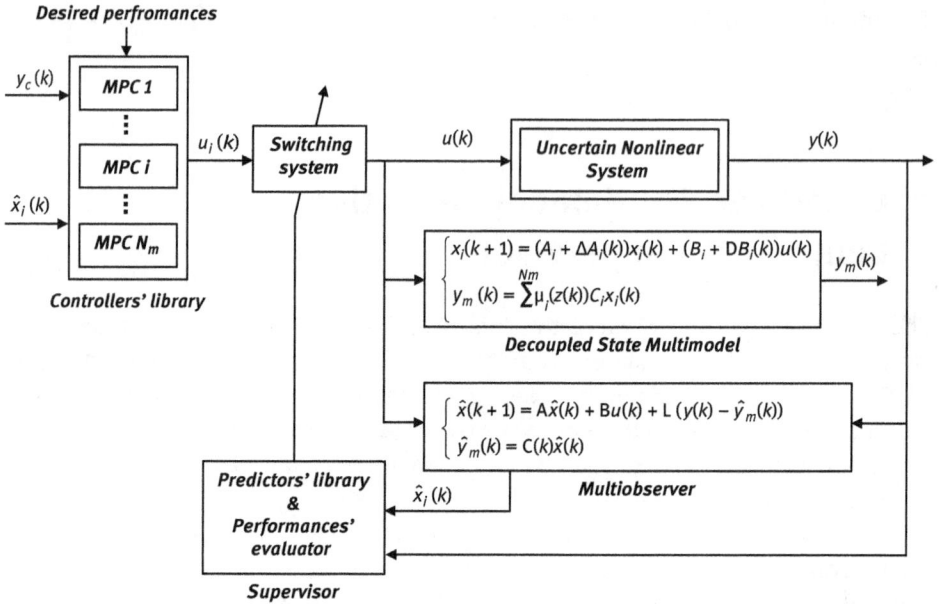

Fig. 1. Supervised decoupled multimodel predictive control based on multi-observer for uncertain nonlinear systems.

$$-\left(\Psi_i^T \vartheta\right)^{-1}\begin{bmatrix} \eta u_i(k-1) \\ 0 \\ \vdots \\ 0 \end{bmatrix} \tag{28}$$

We consider only the first element of $\hat{Y}_{ci}(k)$.

$$\hat{y}_{ci}(k) = \hat{Y}_{ci}(1) \tag{29}$$

5 Simulation Results

Consider the decoupled multimodel with $N_m = 3$ partial models. The numerical matrices' values are given as follows:

Partial model 1:

$$A_1 = \begin{bmatrix} -0.6965 & 0.2218 \\ 0.2430 & 0.0914 \end{bmatrix}, B_1 = \begin{bmatrix} -1.1485 \\ -0.0905 \end{bmatrix}, C_1 = \begin{bmatrix} 1 & 0 \end{bmatrix}$$

$$D_{A1} = \begin{bmatrix} 0.1 & -0.1 \end{bmatrix}, E_{A1} = \begin{bmatrix} -0.2 & 0.2 \end{bmatrix}, D_{B1} = \begin{bmatrix} 1 & -0.1 \end{bmatrix}, E_{B1} = 0.01$$

Partial model 2:

$$A_2 = \begin{bmatrix} -0.5274 & -0.1475 \\ 0.2308 & 0.2425 \end{bmatrix}, B_2 = \begin{bmatrix} -1.3599 \\ 0.2579 \end{bmatrix}, C_2 = \begin{bmatrix} 1 & 0 \end{bmatrix},$$

$$D_{A2} = \begin{bmatrix} 0.1 & -0.01 \end{bmatrix}, E_{A2} = \begin{bmatrix} 0.1 & 0.02 \end{bmatrix} D_{B2} = \begin{bmatrix} 1 & -0.2 \end{bmatrix}, E_{B2} = -0.01$$

Partial model 3:

$$A_3 = \begin{bmatrix} -0.2884 & -0.5253 \\ 0.2000 & 0.1818 \end{bmatrix}, B_3 = \begin{bmatrix} -2.1589 \\ 0.4274 \end{bmatrix}, C_3 = \begin{bmatrix} 1 & 0 \end{bmatrix},$$

$$D_{A3} = \begin{bmatrix} 0.1 & -0.01 \end{bmatrix}, E_{A3} = \begin{bmatrix} 0.2 & 0.02 \end{bmatrix}, D_{B3} = \begin{bmatrix} 2 & -0.1 \end{bmatrix}, E_{B3} = -0.01$$

The application of the designed control strategy based on the designed multi-observer for uncertain systems leads to the following simulation results.

The evolutions of the real and desired outputs are given by Fig. 2.

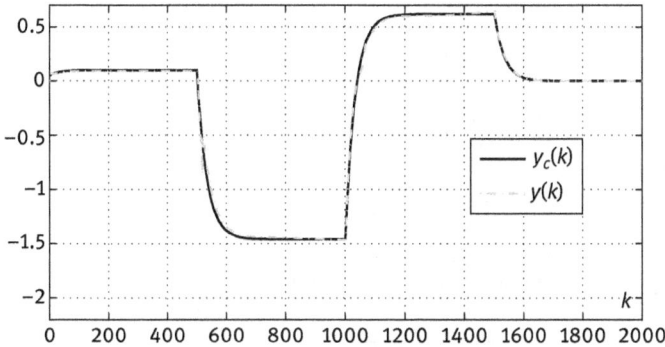

Fig. 2. Evolutions of the real and desired outputs (consideration of uncertainties).

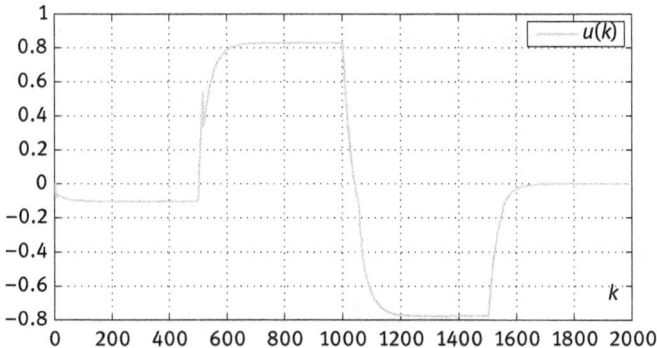

Fig. 3. Evolution of the control signal (consideration of uncertainties).

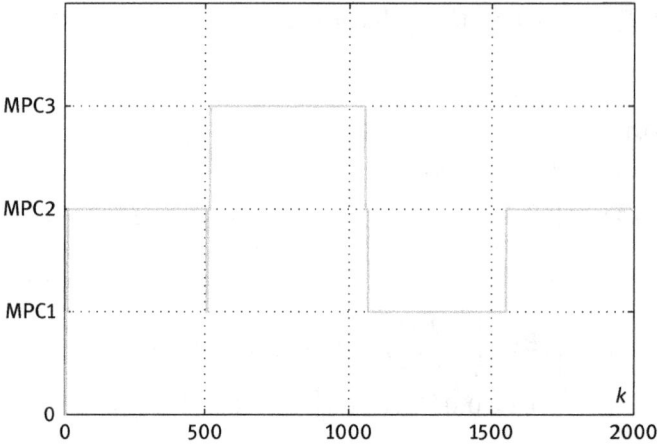

Fig. 4. Commutation between the local controllers (consideration of uncertainties).

(a)

(b)

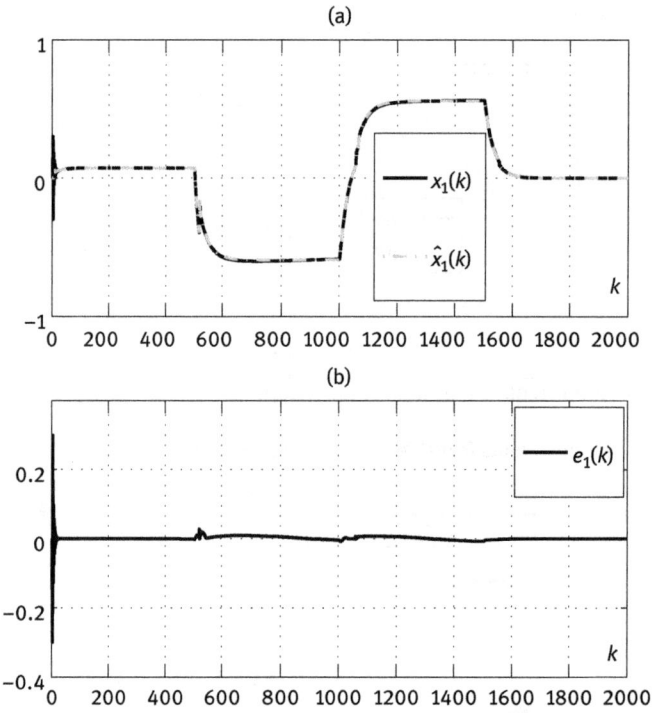

Fig. 5. (a) Evolutions of the state 1 and its estimate (b) Evolution of the estimation error (consideration of uncertainties).

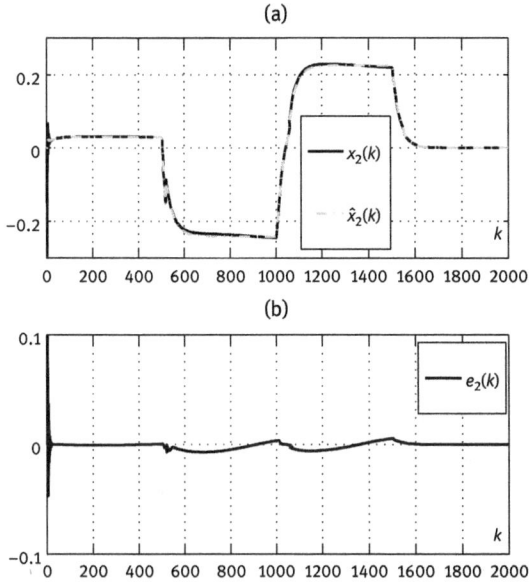

Fig. 6. (a) Evolutions of the state 2 and its estimate (b) Evolution of the estimation error (consideration of uncertainties).

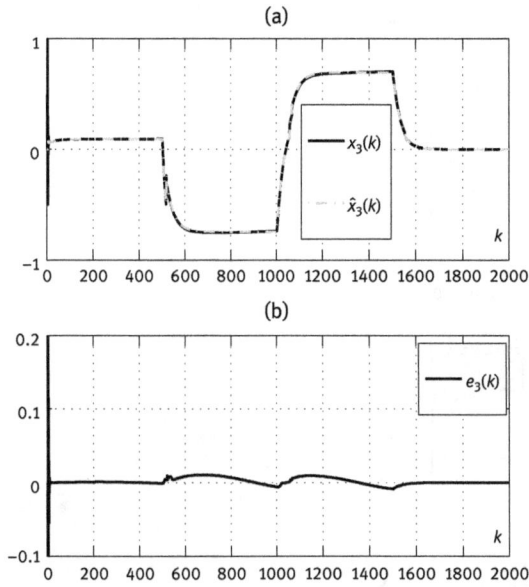

Fig. 7. (a) Evolutions of the state 3 and its estimate (b) Evolution of the estimation error (consideration of uncertainties).

(a)

(b)

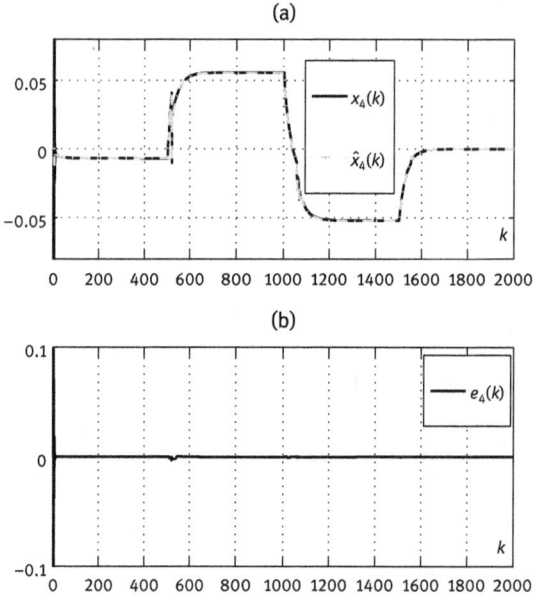

Fig. 8. (a) Evolutions of the state 4 and its estimate (b) Evolution of the estimation error (consideration of uncertainties).

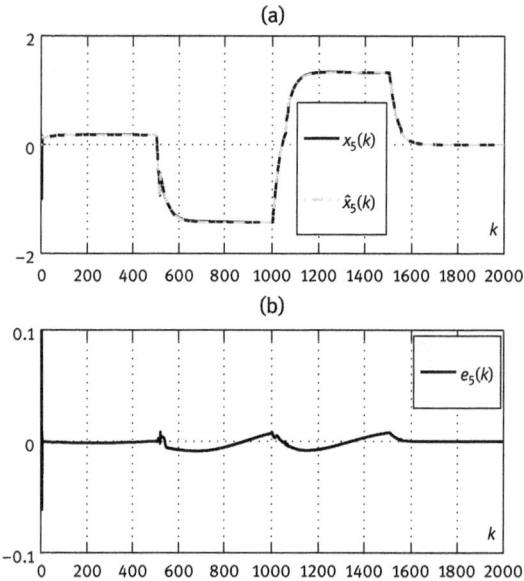

(a)

(b)

Fig. 9. (a) Evolutions of the state 5 and its estimate (b) Evolution of the estimation error (consideration of uncertainties).

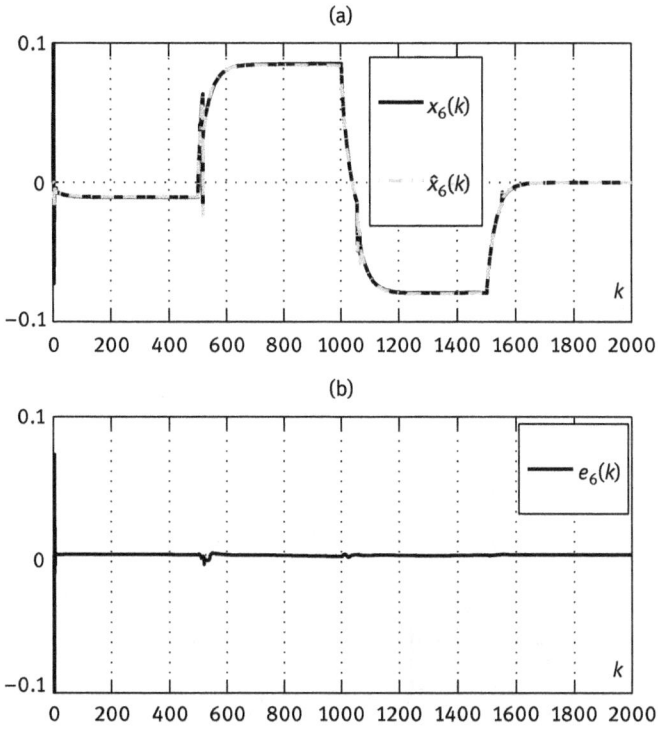

Fig. 10. (a) Evolutions of the state 6 and its estimate (b) Evolution of the estimation error (consideration of uncertainties).

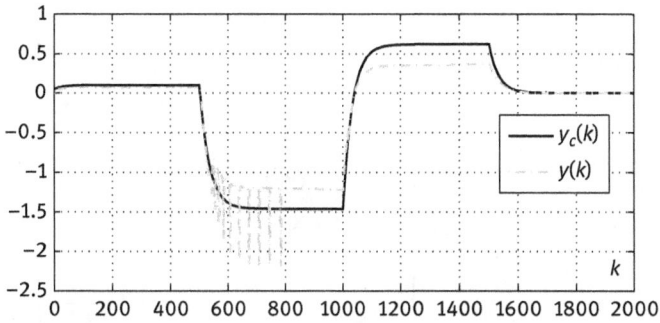

Fig. 11. Evolutions of the real and desired outputs (non-consideration of uncertainties).

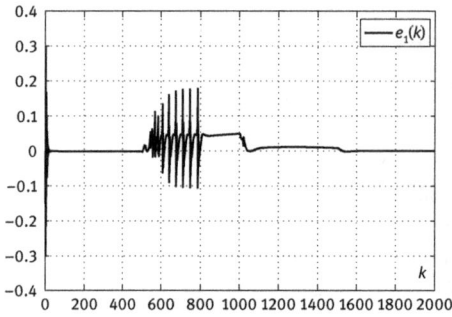

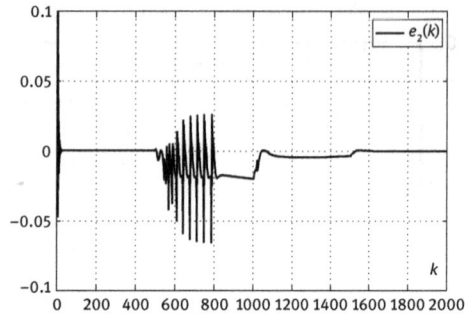

Fig. 12. Evolution of the state estimation errors of the partial model 1 (non-consideration of uncertainties).

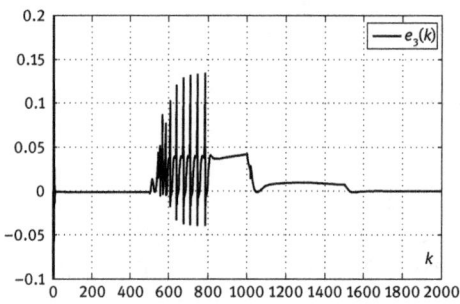

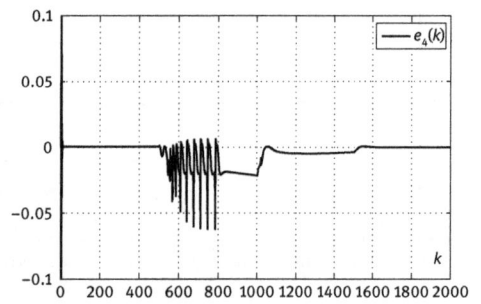

Fig. 13. Evolution of the state estimation errors of the partial model 2 (non-consideration of uncertainties).

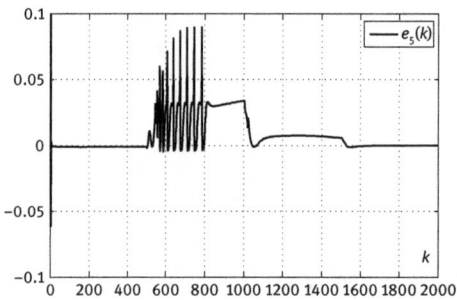

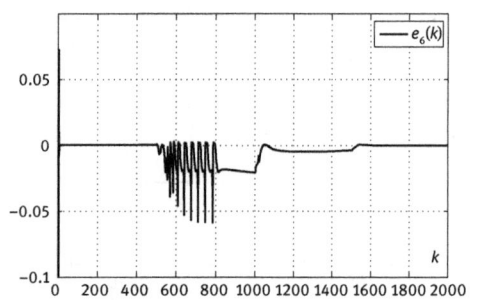

Fig. 14. Evolution of the state estimation errors of the partial model 3 (non-consideration of uncertainties).

Figure 3 illustrates the evolution of the control signal.

Figure 4 shows the commutation between partial controllers.

The evolutions of the real and estimate states as well as the estimation errors are shown in the Figs. 5–10.

These figures illustrate the efficiency of the decoupled multimodel predictive control for discrete-time uncertain nonlinear systems when a multi-observer for the state estimation is used. This control strategy provides good closed-loop performances when state variables are accurately estimated.

Now, we assume that the real system contains parametric uncertainties. However, these uncertainties are ignored in the modeling phase. So, the obtained multimodel is no longer a faithful representation of the real system. Consequently, the uncertainties are not considered in the multi-observer synthesis. The effect of the non consideration of uncertainties on the quality of state estimation and subsequently on the efficiency of the control law is illustrated by the following figures.

Figure 11 illustrates the evolutions of the real and desired outputs.

The evolutions of the estimation errors are given by Figs. 12–14.

It can be easily seen that the non consideration of the uncertainties in the synthesis of the multi-observer has bad effects on the state estimation quality, and since the control law depends on these estimates (equation 26), we can conclude that the degradation of control law performance is due to the bad estimation quality. This limit is already overcame due to the designed multi-observer that takes into account the system uncertainties.

6 Conclusion

In this paper, we have combined the multi-observer for the state estimation of discrete-time uncertain decoupled state multimodel with the multimodel predictive control in order to obtain an efficient control strategy for uncertain discrete-time nonlinear systems. The designed multi-observer provides the estimates of the states required for the elaboration of an efficient control law. A supervisor including a performance criterion is exploited for the selection of the adequate partial controller that satisfies the closed-loop objectives. A simulation example is carried out to show the significance of this control strategy that shows good closed-loop performances.

Bibliography

[1]   M. Abbaszadeh and HJ. Marquez. Lmi optimization approach to robust H_∞ observer design and static output feedback stabilization for discrete-time nonlinearuncertain systems. *International Journal Of Robust and Nonlinear Control*, 19:313–340, 2009.

[2]   J. Abonyi, R. Babuska, and F. Szeifert. Modified gath-geva fuzzy clustering for identification of takagi-sugeno fuzzy models. *IEEE Transactions on Systems, Man and Cybernetics, Part B: Cybernetics*, 32(5):612–621, 2002.

[3]   R. Aguilar-Lopez and R. Maya-Yescas. State estimation for nonlinear systems under model uncertainties: a class of sliding-mode observers. *Journal of Process Control*, 15(3):363–370, 2005.

[4]   X.B. Hu and W.H. Chen. Model predictive control for constrained systems with uncertain state-delays. *International Journal of Robust and Nonlinear Control*, 14(17):1421–1432, 2004.

[5]   S. Ibrir and S. Diop. Novel lmi conditions for observer-based stabilization of lipschitzian nonlinear systems and uncertain linear systems in discrete-time. *Applied Mathematics and Computation*, 206(2):579–588, 2008.

[6]   A.M. Nagy Kiss. *Analyse et synthèse de multimodèles pour le diagnostic. Application à une station d'épuration*. PhD thesis, Institut National Polytechnique de Lorraine, France, 2010.

[7]   A. Messaoud, M. Ltaief, and R. Ben Abdennour. Supervision based on partial predictors for a multimodel generalized predictive control: experimental validation on a semi-batch reactor. *International Journal of Modeling, Identification and Control*, 6(4):333–340, 2009.

[8]   A. Messaoud, M. Ltaief, and R. Ben Abdennour. Supervision based on a multipredictor for an uncoupled state multimodel predictive control. In *The 6th International Conference on Electrical Systems and Automatic Control*, 2010.

[9]   W. El Messoussi, O. Pagès, and A. El Hajjaji. Robust pole placement for fuzzy models with parametric uncertainties: An lmi approach. In *Proceedings of the 4th Eusflat and 11th LFA Congress*, pages 810–815, 2005.

[10] W. El Messoussi, O. Pagès, and A. El Hajjaji. Observer-based robust control of uncertain fuzzy dynamic systems with pole placement constraints: An lmi approach. In *Proceedings of the IEEE American Control Conference*, pages 2203–2208, 2006.

[11] H.N. Nounou and K.M. Passino. Fuzzy model predictive control: Techniques, stability issues, and examples. *Proceedings of the IEEE International Symposium on Intelligent Control*, pages 423–428, 1999.

[12] R. Orjuela. *Contribution à l'estimation d'état et au diagnostic des systèmes représentés par des multimodèles*. PhD thesis, Institut National Polytechnique de Lorraine, France, 2008.

[13] R. Orjuela, D. Maquin, and J. Ragot. Nonlinear system identification using uncoupled state multiple-model approach. In *4th Workshop on Advanced Control and Diagnosis, ACD'20–06, Nancy, France*, 2006.

[14] R. Orjuela, B. Marx, D. Maquin, and J. Ragot. State estimation of nonlinear discrete-time systems based on the decoupled multiple model approach. In *4th International Conference on Informatics in Control, Automation and Robotics, ICINCO 2007, Angers, France*, 2007.

[15] R. Orjuela, B. Marx, J. Ragot, and D. Maquin. Conception d'observateurs robustes pour des systèmes non linéaires incertains: une stratégie multimodèle. In *5eme Conférence Internationale Francophone d'Automatique, Bucarest, Roumanie*, 2008.

[16] R. Orjuela, B. Marx, J. Ragot, and D. Maquin. Design of robust h_∞ observers for nonlinear systems using a multiple model. In *17th IFAC World Congress, Séoul, Corée du Sud*, 2008.

[17] O. Pagès, C. Bernard, O. Raul, and M. Pascal. Control system design by using a multi-controller approach with a real-time experimentation for a robot wrist. *International Journal Of Control*, 75(16):1321–1334, 2002.

[18] H. Sarimveis and G. Bafas. Fuzzy model predictive control of non-linear processes using genetic algorithms. *Fuzzy Sets and Systems*, 139(1):59–80, 2003.

[19] C.L. Su and S.Q. Wang. Robust model predictive control for discrete uncertain nonlinear systems with time-delay via fuzzy model. *Journal of Zhejiang University SCIENCE A*, 7(10):1723–1732, 2006.

[20] K. Tanaka and M. Sano. A robust stabilization problem of fuzzy control systems and its application to backing up control of a truck-trailer. *IEEE Transactions on fuzzy systems*, 2(2):119–134, 1994.
[21] Y. Zhang, M. Liu, W. Jianhui, S. Gu, and G. Li. Robust model predictive control for uncertain discrete-time system with both states and input delays. In *Chinese Control and Decision Conference*, number 279 - 284, 2008.

Biographies

Samah Ben Atia received her Engineering Diploma in Electric-Automatic engineering, in 2011, and the Master degree in Automatic Control and Intelligent Techniques, in 2012, from National School of Engineers of Gabes - Tunisia. Currently, she is pursuing her PhD thesis at CONPRI (Research Unit of Numerical Control of Industrial Processes at ENIG). Her areas of interest include multimodel and multicontrol approaches for uncertain and time-delay nonlinear systems.

Anis Messaoud received the Engineering and Master degrees in electrical Engineers and automatic control from National School of Engineers of Gabes-Tunisia, in 2004 and 2006 respectively. In 2010, he obtained his Ph. D. degree in electrical-automatic engineering from the National School of Engineers of Gabes, Tunisia. His specific research interests are in the area predictive control of complex systems, Multimodel and Multicontrol approaches. He is currently an associate professor in the Electric Engineering Department in National School of Engineers of Gabes-Tunisia.

Ridha Ben Abdennour received the Doctorat de Spécialité degree from Higher School of Technical Education in 1987, and the Doctorat d'Etat degree from the National School of Engineers of Tunis-Tunisia, in 1996. He is Professor in Automatic Control at the National School of Engineers of Gabes - Tunisia. He was chairman of the Electrical Engineering Department and the Director of the High Institute of Technological Studies of Gabes. Ridha BEN ABDENNOUR is the Head of the Research Unit of Numerical Control of Industrial Processes and he is the founder and the honorary President of the Tunisian Association of Automatic Control and Digitalization. His research is on Identification, Multimodel & Multicontrol approaches, Numerical Control and Supervision of Industrial Processes. He is the co-author of a book on Identification and Numerical Control of Industrial Processes and he is the author of more than 300 publications. Ridha BEN ABDENNOUR participated in the organization of many Conferences and he was member of some scientific committees of congresses.

M. Allaoui, A. Messaoud and R. Ben Abdennour

Robustness Enhancement of Proportional *Q*-integral Multiobserver in the Case of Non-stationary Sinusoidal Unknown Inputs

Abstract: This paper deals with the robustness enhancement of proportional *Q*-integral multiobserver in the case of non-stationary sinusoidal unknown inputs. The multimodel approach is proposed in order to overcome the complexity problems of nonlinear systems. The proposed multiobserver uses the multi-integral strategy in order to provide a simultaneous estimation of the state and unknown inputs. A new robust strategy allowing the minimization of the non-stationary sinusoidal unknown inputs impact on the estimation error is developed. The significance of the proposed study is illustrated via a simulation example.

Keywords: Nonlinear systems, decoupled state multimodel, non-stationary sinusoidal unknown inputs, multiobserver, stability.

1 Introduction

In the automation field, the accuracy of the information on the state variables still represents a fundamental tool that served control theory [3, 13, 14]. Indeed, control systems require always accurate information on the processes internal states in order to generate control laws ensuring that the process behavior follows a desired behavior for future instants [6, 12]. Thus, it can be noted that the control loops performances are directly related to the accuracy quality of the information on the state variables [12, 13].

Thereafter, it is noteworthy that physical and economic restrictions prevent direct measurements of state variables and even makes them impossible. To this end, the use of the observers constitutes a widely adopted solution to provide accurate state estimation [1, 4, 8, 16].

Nevertheless, the observers design still requires a dynamic model that accurately reflects the process behavior. Thus, the estimation quality is proportionally tied to the ability of the considered model to represent the process behavior.

M. Allaoui, A. Messaoud and R. Ben Abdennour: Research Unit: Numerical Control of Industrial Processes (CONPRI), University de Gabès, Ecole Nationale d'Ingénieurs de Gabès, Université de Gabès, Tunisie, M. Allaoui, email: allaoui.mouhib@gmail.com, A. Messaoud, email: messaoud_anis@yahoo.fr, R. Ben Abdennour, email: ridha.benabdennour@enig.rnu.tn.

De Gruyter Oldenbourg, ASSD – Advances in Systems, Signals and Devices, Volume 5, 2018, pp. 55–76.
DOI 10.1515/9783110470468-004

For this reason, it is wise to take into account the external phenomena, result of process interaction with its environment, in the considered dynamic model [3, 15, 18]. Indeed, since the system cannot be isolated from its environment, it is well known that different external dynamics, namely the unknown inputs and disturbances, can intervene on the dynamics of the physical system. The non-consideration of these nonmeasurable inputs leads to performance degradation of the representative model and makes the state estimation biased and much more difficult.

For this purpose, state and unknown input estimation has been extensively investigated and considerable attention has been drawn to the unknown input observers for linear and nonlinear systems [1, 5, 8–10, 17, 20, 21]. This observer undergoes structural changes compared to classical observers allowing it to take into account these unknown inputs and provide robust and accurate estimates [4, 7].

Currently, several studies on the synthesis and design of unknown input observers for linear systems give it a certain maturity [4, 7, 8, 20, 21]. However, it is well known that linear models are not able to represent the behavior of physical systems that around an operating point. In fact, most physical systems have a much more complex nonlinear behavior to represent through a linear model in a large operating space [9]. Moreover, the consideration of nonlinear models representation deals with several challenges. Indeed, the complexity of the adapted model representation makes the synthesis of observers very difficult [3, 9]. In addition, any approximation of the nonlinear characterization to a defined class of nonlinearity can lead to performance degradation as soon as the simplifying assumptions become invalid.

To overcome these problems the multimodel approach can be an interesting alternative [1, 5, 6, 10, 13, 14, 17, 18]. Thereafter, the multimodel approach constitutes a very interesting mathematical representation. It is used to represent the behavior of the nonlinear systems, whatever their complexity, by a simple structure based on linear partial models interpolated by weighting functions [2, 11, 12, 16, 19].

This work deals with a proportional Q-integral multiobserver for discrete-time nonlinear systems described by a decoupled state multimodel. This proposed multiobserver is designed to decouple the state estimation error from unknown inputs which leads to accurate estimation of both states and non-stationary sinusoidal unknown inputs.

The outline of this paper is organized as follows. In the second section, the decoupled state multimodel structure for the representation of nonlinear systems is presented. The third section is devoted to the design of the proportional Q-integral multiobserver. The fourth part illustrates the significance of the proposed robustness enhancement strategy via a simulation example. A conclusion achieves this paper.

2 Decoupled State Multimodel Structure for the Representation of Nonlinear systems

Consider a discrete-time non linear system described by the following decoupled state multimodel [16] and [18]:

$$
\begin{cases}
x_i(k+1) &= A_i x_i(k) + B_i u(k) + E_{e_i} P_s(k) \\
y_i(k) &= C_i x_i(k) + E_{s_i} P_s(k) \\
y_{MM}(k) &= \sum_{i=1}^{N_m} \mu_i[v(k)] y_i(k)
\end{cases}
\tag{1}
$$

where:

$x_i(k) \in \mathbb{R}^{n_i}$ denotes the state vector of the i–th partial model.

$y_i(k)$ and $y_{MM}(k) \in \mathbb{R}^p$ are respectively the output of the i–th partial model and the multimodel output vectors.

$u(k) \in \mathbb{R}^m$ is the input vector.

$P_s(k) \in \mathbb{R}^l$ is the non-stationary sinusoidal unknown inputs characterized by its magnitude P_m and its period T_s (the multiple integer of the sampling period).

N_m is the number of partial models.

The matrices $A_i \in \mathbb{R}^{n_i \times n_i}$, $B_i \in \mathbb{R}^{n_i \times m}$, $C_i \in \mathbb{R}^{p \times n_i}$, $E_{e_i} \in \mathbb{R}^{n_i \times l}$ and $E_{s_i} \in \mathbb{R}^{p \times l}$ are known and appropriately dimensioned.

The breakdown of the nonlinear system operating space is performed via the decision variable $v(k)$. It is assumed to be measurable and available in real time.

Thereafter, we associate to each operating zone a linear partial model.

The overlapping of these linear models, using weighting functions, ensures a nonlinear dynamic behavior faithful to that of the real system. It should be mentioned that these weighting functions depend on the decision variables.

They are usually selected to satisfy the following properties:

$$
\begin{cases}
\sum_{i=1}^{N_m} \mu_i[v(k)] = 1 \\
0 \leq \mu_i[v(k)] \leq 1,
\end{cases}
\quad \forall\, i = 1, \ldots, N_m, \forall k
\tag{2}
$$

Therefore, thanks to these properties, the contribution of the set of the linear partial model can be considered simultaneously. This provides to the multimodel a real nonlinear dynamic behavior rather than a piecewise linear behavior.

Generally these weighting functions can be represented by the normalized Gaussian functions described as below:

$$\mu_i[v(k)] = \frac{e^{\left(-\frac{v(k)-c_i}{\sigma_d}\right)^2}}{\sum\limits_{j=1}^{N_m} e^{\left(-\frac{v(k)-c_j}{\sigma_d}\right)^2}}, \qquad i = 1, 2, \ldots, N_m \tag{3}$$

where:
c_i $(i = 1, \ldots, N_m)$ are the centers and σ_d is the dispersion.

Within a perspective of ensuring the global convergence of the estimation error, it is wise to consider mixed outputs of partial models during the design of the observer. Indeed, it is worth mentioning that the interpolation of a set of stable partial models may cause unstable multi-model and conversely. Same for observers, the design of independent observers for each partial model can lead to an unstable global estimation error. To overcome this problem, we propose to define the following augmented vector:

$$x_{cf}(k) = \begin{bmatrix} x_1^T(k) & \cdots & x_i^T(k) & \cdots & x_{N_m}^T(k) \end{bmatrix}^T \in \mathbb{R}^n, \qquad n = \sum_{i=1}^{N_m} n_i \tag{4}$$

Subsequently, a compact structure of the decoupled state multimodel (1) can be written as follows:

$$\begin{cases} x_{cf}(k+1) &= A_{cf}x_{cf}(k) + B_{cf}u(k) + E_{e_{cf}}P_s(k) \\ y_{MM}(k) &= C_{cf}x_{cf}(k) + E_{s_{cf}}P_s(k) \end{cases} \tag{5}$$

where:

$$A_{cf} = \begin{bmatrix} A_1 & 0 & \cdots & & 0 \\ 0 & \ddots & & & \vdots \\ \vdots & & A_i & & \\ & & & \ddots & 0 \\ 0 & \cdots & & 0 & A_{N_m} \end{bmatrix} \in \mathbb{R}^{n \times n}, \quad B_{cf} = \begin{bmatrix} B_1 \\ \vdots \\ B_i \\ \vdots \\ B_{N_m} \end{bmatrix} \in \mathbb{R}^{n \times m}$$

$$C_{cf}(k) = \begin{bmatrix} \mu_1[v(k)]C_1 \ldots \mu_i[v(k)]C_i \ldots \mu_{N_m}[v(k)]C_{N_m} \end{bmatrix} \in \mathbb{R}^{p \times n}$$

$$E_{e_{cf}} = \begin{bmatrix} E_{e_1}^T & \cdots & E_{e_i}^T & \cdots & E_{e_{N_m}}^T \end{bmatrix}^T \in \mathbb{R}^{n \times l}, \quad E_{s_{cf}}(k) = \sum_{i=1}^{N_m} \mu_i[v(k)]E_{s_i}$$

Consider the following bloc matrix, $\tilde{C}_{cf_i} = \begin{bmatrix} 0 & \cdots & C_i & \cdots & 0 \end{bmatrix} \in \mathbb{R}^{p \times n}$, and reminding the properties of the convex sum (2), the matrix $C_{cf}(k)$ can be expressed as follows:

$$C_{cf}(k) = \sum_{i=1}^{N_m} \mu_i[v(k)]\tilde{C}_{cf_i} \tag{6}$$

3 Proportional Q-Integral Multiobserver Design for Decoupled State Multimodel

3.1 Multiobserver Structure

In view of providing simultaneous estimation of the state and the unknown inputs using decoupled state multimodel (1), the proportional Q-integral multiobserver described below is exploited [1, 16] and [18].

Thus, the reconstruction of the state variables of the multimodel is performed by the proportional multiobserver (7), however the use of integral actions (8) allows the reconstruction of the unknown inputs.

$$
\begin{cases}
\hat{x}_i(k+1) = A_i\hat{x}_i(k) + B_iu(k) + E_{e_i}\hat{P}_{s_0}(k) + K_i\left[y(k) - \hat{y}_{MM}(k)\right] \\
\hat{y}_i(k) = C_i\hat{x}_i(k) + E_{s_i}\hat{P}_{s_0}(k) \\
\hat{y}_{MM}(k) = \sum_{i=1}^{N_m} \mu_i[v_i(k)]\hat{y}_i(k)
\end{cases}
\tag{7}
$$

$$
\begin{cases}
\hat{P}_{s_q}(k+1) = \hat{P}_{s_q}(k) + K_q\left[y(k) - \hat{y}_{MM}(k)\right] + \hat{P}_{s_{q+1}}(k), \quad q \in [0, Q-1] \\
\hat{P}_{s_Q}(k+1) = \hat{P}_{s_Q}(k) + K_Q\left[y(k) - \hat{y}_{MM}(k)\right]
\end{cases}
\tag{8}
$$

where:

$\hat{x}_i(k) \in \mathbb{R}^{n_i}$ is the ith partial model state estimation vector.

$\hat{y}_i(k), \hat{y}_{MM}(k), y(k) \in \mathbb{R}^p$ denote respectively the ith partial model output estimation vector $y_i(k)$, the multimodel output estimation vector $y_{MM}(k)$ and the measured output vector.

$\hat{P}_{s_0}(k) \in \mathbb{R}^l$ is the non-stationary sinusoidal unknown inputs estimation.

$K_i \in \mathbb{R}^{n_i \times p}$ represents the gain of the i−th partial model.

$K_q \in \mathbb{R}^{l \times p}$ is the gain of q−integral observer ($q \in [0, Q]$).

In order to rewrite the proportional multiobserver (7) in a compact form, let us define the following vector:

$$
\hat{x}_{cf}(k) = \left[\hat{x}_1^T(k) \quad \cdots \quad \hat{x}_i^T(k) \quad \cdots \quad \hat{x}_{N_m}^T(k)\right]^T \in \mathbb{R}^n
\tag{9}
$$

Thus, the proportional multiobserver can be rewritten as follows:

$$
\begin{cases}
\hat{x}_{cf}(k+1) = A_{cf}\hat{x}_{cf}(k) + B_{cf}u(k) + E_{e_{cf}}\hat{P}_{s_0}(k) + K_{p_{cf}}\left[y(k) - \hat{y}_{MM}(k)\right] \\
\hat{y}_{MM}(k) = C_{cf}(k)\hat{x}_{cf}(k) + E_{s_{cf}}(k)\hat{P}_{s_0}(k)
\end{cases}
\tag{10}
$$

where $\hat{x}_{cf}(k) \in \mathbb{R}^n$ is the estimated of $x_{cf}(k)$. $K_{p_{cf}} \in \mathbb{R}^{l \times n}$ is the gain of the augmented proportional multiobserver.

3.2 Multiobserver Synthesis

This section deals with the estimation problem, based on the decoupled state multimodel (4) and using the proportional Q–integral multiobserver described by (10) and (8).

Within this framework, in order to study the convergence of the state and the unknown inputs estimation errors, it is necessary to establish their expressions. For this purpose, we propose firstly to define the q–th-difference operator of a signal $P_s(k)$:

$$\Delta^{(q)}P_s(k) = \Delta^{(q-1)}(\Delta P_s(k)) = \Delta(\Delta^{(q-1)}P_s(k))$$

Furthermore, the state and the non-stationary sinusoidal unknown inputs estimation errors are defined by:

$$\begin{cases} e_x(k) = x_{if}(k) - \hat{x}_{if}(k) \\ e_{p_q}(k) = \Delta^{(q)}P_s(k) - \hat{P}_{s_q}(k) \quad q = 0 \ldots Q \end{cases} \tag{11}$$

where: $\Delta^{(0)}P_s(k) = P_s(k)$.

Thereafter, substituting the expressions of $x_{cf}(k)$, $\hat{x}_{cf}(k)$ given respectively by (5) and (10) in the equation of $e_x(k)$, the dynamics of the state estimation error is expressed as follows:

$$e_x(k+1) = [A_{cf} - K_{p_{cf}}C_{cf}(k)]e_x(k) + [E_{e_{cf}} - K_{p_{cf}}E_{s_{cf}}(k)]e_{p_0}(k) \tag{12}$$

Other hand, using the equations (8) and (11) the dynamics of the non-stationary sinusoidal unknown inputs estimation error is described by:

$$e_{p_q}(k) = -K_q C_{cf}(k)e_x(k) - K_q E_{s_{cf}}(k)e_{p_0}(k) + e_{p_q}(k) + e_{p_{q+1}}(k), \text{ for } q \in [0, Q-1] \tag{13}$$

$$e_{p_Q}(k+1) = -K_Q C_{cf}(k)e_x(k) - K_Q E_{s_{cf}}(k)e_{p_0}(k) + e_{p_Q}(k) + \Delta^{(Q+1)}P_s(k) \tag{14}$$

In order to make the representation of the dynamics of the state and the non-stationary sinusoidal unknown inputs estimation errors and their differences in a compact form, we propose to define the following augmented error vector:

$$\psi(k) = \begin{bmatrix} e_x^T(k) & e_{p_0}^T(k) & \cdots & e_{p_q}^T(k) & \cdots & e_{p_Q}^T(k) \end{bmatrix}^T \in \mathbb{R}^\vartheta, \; \vartheta = n + l(q+1) \tag{15}$$

Using the previous augmented error vector and taking everything into account, the dynamic of the error vector $\psi(k+1)$ is written as follows:

$$\psi(k+1) = [\Omega - K_r\theta(k)]\psi(k) + \Pi\Delta^{(Q+1)}P_s(k) \tag{16}$$

where:

$$\Omega = \begin{bmatrix} A_{cf} & E_{e_{cf}} & 0 & \cdots & & 0 \\ 0 & I_l & I_l & 0 & & 0 \\ & \ddots & \ddots & \ddots & \ddots & \vdots \\ \vdots & & & \ddots & \ddots & 0 \\ & & & \ddots & I_l & I_l \\ 0 & & \cdots & & 0 & I_l \end{bmatrix} \in \mathbb{R}^{\vartheta \times \vartheta},$$

$$K_r = \begin{bmatrix} K_{p_{cf}}^T & K_0^T & \cdots & K_Q^T \end{bmatrix} \in \mathbb{R}^{\vartheta \times p}$$

$$\theta(k) = \begin{bmatrix} C_{cf}(k) & E_{s_{cf}}(k) & 0 & \cdots & 0 \end{bmatrix} \in \mathbb{R}^{p \times \vartheta}$$

$$\Pi = \begin{bmatrix} 0 & \cdots & 0 & I_l \end{bmatrix}^T \in \mathbb{R}^{\vartheta \times l}$$

Making use of the properties of the convex sum and the expression described by (6), we can write:

$$\varphi(k) = \sum_{i=1}^{N_m} \mu_i[v(k)] \, [\Omega - K_r \theta_i] \tag{17}$$

where:

$$\theta_i = \begin{bmatrix} \tilde{C}_{i_{cf}} & E_{s_i} & 0 & \cdots & 0 \end{bmatrix}$$

Thereby, exploiting the previous equation the expression of the dynamic of the augmented error (16) becomes:

$$\psi(k+1) = \varphi(k)\psi(k) + \Pi\Delta^{(Q+1)}P_s(k) \tag{18}$$

Examining the equation (18), we can note that the estimation error is directly affected by the $(Q+1)^{th}$ difference of the unknown input.

In this framework, for the purpose of decoupling the estimation error with respect to unknown inputs several attempts were recorded. Indeed, concerning polynomial unknown input with d_c known degree, the perfect decoupling can be achieved by employing $d_c + 1$ integral actions [16, 18].

Within the same framework, when the unknown input model does not take an exact polynomial form, the proposed condition can not provide a complete decoupling between the unknown input and the estimation error $\psi(k)$. Indeed, the consideration of d_c+1 integral actions allows only to decouple the influence of the d_c degrees' polynomial part. Thereafter, the $(d_c + 1)$–th difference of the unknown input $\Delta^{(d_c+1)}P_s(k)$ is non-nul. Thus, the estimated state and unknown input remaining biased and can not be perfectly accomplished. To remedy this problem, Orjuela proposed to attenuate the non-nul part of the unknown inputs' $(d_c + 1)$–th difference provided it is bounded [18].

Assumption 1: The unknown input must satisfy the following condition:

$$\Delta^{(Q+1)} P_s(k) = \sigma_a(k), \text{ where } \|\sigma_a(k)\|_2^2 < \infty \tag{19}$$

Thereafter, the equation characterizing the estimation error (18) can be rewritten as follows:

$$\psi(k+1) = \varphi(k)\psi(k) + \Pi\sigma_a(k) \tag{20}$$

Considering the equation (20), we note that the dynamics of the state estimation error is biased by $\sigma_a(k)$ which causes the degradation of the state estimation's quality and prevents the estimation error to converge towards zero.

Thus, we must take into account $\sigma_a(k)$ in the design of the multiobserver. Thereafter the objective signal to be attenuated is considered:

$$O_s(k) = M\psi(k) \tag{21}$$

where: M is a matrix of appropriate dimensions. It allows to reduce the impact of $\sigma_a(k)$ on the components of the vector $\psi(k)$.

Thereafter, the constraint to be satisfied during the synthesis of the multiobserver can be expressed by:

✓ For $\sigma_a(k) \neq 0$ and $O_s(0) = 0$

$$\|O_s(k)\|_2^2 < \gamma^2 \|\sigma_a(k)\|_2^2 \tag{22}$$

where: γ denotes the $\mathcal{L}_2$ gain to be minimized. It indicates the level of attenuation between $\sigma_a(k)$ and $O_s(k)$.

The estimation error should tend exponentially to zero if no $\sigma_a(k)$ acting on the system and must ensure the performance requirement described by (22). The convergence conditions are represented in the following theorem [16, 17] and [18]:

Theorem 1:
Consider the Assumption 1, the estimation error between the decoupled state multimodel (5) and the proportional Q–integral multiobservers (8) and (10), satisfying the requirement (22) and tend exponentially to zero (if σ_a (k)=0), if there exist two matrices $X = X^T > 0$ and W of appropriate dimensions and a scalar $\bar{\gamma} > 0$ solutions of the following convex optimization problem:

$$\min \bar{\gamma}$$
under the constraints :
$$\begin{bmatrix} M^T M - (1 - 2\alpha)X & 0 & (X\Omega - W\theta_i)^T \\ 0 & -\bar{\gamma}I & (P\Pi)^T \\ (X\Omega - W\theta_i) & (P\Pi) & -X \end{bmatrix} < 0, \text{ for } i = 1\ldots N_m \tag{23}$$

If this inequality has a solution for a given rate decay,

$$0 < \alpha < 0.5$$

the multiobserver gain is determined by:

$$K_r = X^{-1} W \tag{24}$$

and the $\mathcal{L}_2$ gain γ is given by:

$$\gamma = \sqrt{\bar{\gamma}} \tag{25}$$

3.3 Simulation Results

Consider the discrete-time nonlinear system modeled by a decoupled state multimodel. This multimodel is defined by two heterogeneous partial models described by:

Partial model 1:

$$A_1 = \begin{bmatrix} -0.32 & -0.04 \\ 1 & 0 \end{bmatrix}, B_1 = \begin{bmatrix} 1 \\ 0 \end{bmatrix}, C_1 = \begin{bmatrix} 0 \\ 0.3 \end{bmatrix}^T, E_{e_1} = \begin{bmatrix} 0.2 \\ 0.1 \end{bmatrix}, E_{s_1} = 0.4$$

Partial model 2:

$$A_2 = \begin{bmatrix} -0.15 & 0.45 & 0.3 \\ -0.1 & -0.7 & -0.2 \\ -0.2 & 0.3 & -0.8 \end{bmatrix}, B_2 = \begin{bmatrix} 0.5 \\ 0.2 \\ 0.1 \end{bmatrix}, C_2 = \begin{bmatrix} 1 \\ 0 \\ 0 \end{bmatrix}^T, E_{e_2} = \begin{bmatrix} 0.1 \\ 0.2 \\ 0.3 \end{bmatrix}, E_{s_2} = 0.5$$

The evolutions of the input is depicted in Fig. 1.

The weighting functions μ_i depend on the decision variable $v(k)$ that is assumed to be the input signal $u(k) \in [0, 1]$.

Thereafter, the values of dispersion and centers are given as following:

$$\sigma_d = 0.2, c_1 = 0.25 \text{ and } c_2 = 0.75$$

The unknown input is defined by:

$$P_s(k) = 0.25 + P_m(k) \sin\left(\frac{2k\pi}{T_s(k)}\right) \tag{26}$$

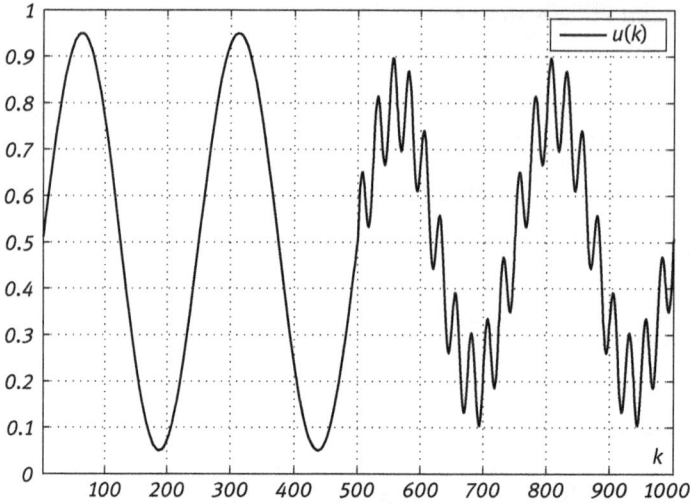

Fig. 1. Evolution of the input.

where:

$$P_m(k) = \begin{cases} 0.15 & if & 1 \le k < 300 \quad or \quad k \ge 600 \\ 0.2 & if & 300 \le k < 600 \end{cases}$$

$$T_s(k) = \begin{cases} 100 & if & 1 \le k < 300 \\ 75 & if & 300 \le k < 600 \\ 200 & else \end{cases}$$

Figure 2 shows the evolution of the unknown input.

The considered unknown input consists of two different forms, a non-stationary sinusoidal part characterized by its maximum magnitude $P_m = 0.2$ and its minimum period $T_s = 75$, and another polynomial part with known degree $d_c = 0$.

Based on the proposed classical strategy, the number of integral actions necessary to decouple the polynomial part is equal to $Q = d_c + 1 = 1$.

By considering the decay rate $\alpha = 0.1$ and $M = [I_{(5\times5)} \; 0_{(5\times3)}]$ the optimal solutions satisfying conditions of Theorem 1 are fulfilled with:

$$k_r = [\; 0.2940 \quad 0.2933 \quad 0.0793 \quad 0.3716 \quad 0.3378 \quad 1.6875 \;]^T \tag{27}$$

with a minimal attenuation level given by $\gamma = 0.8857$.

The evolution of $\sigma_a(k)$ is depicted in Fig. 3:

The unknown input and the multimodel output estimation errors are shown in the following figures:

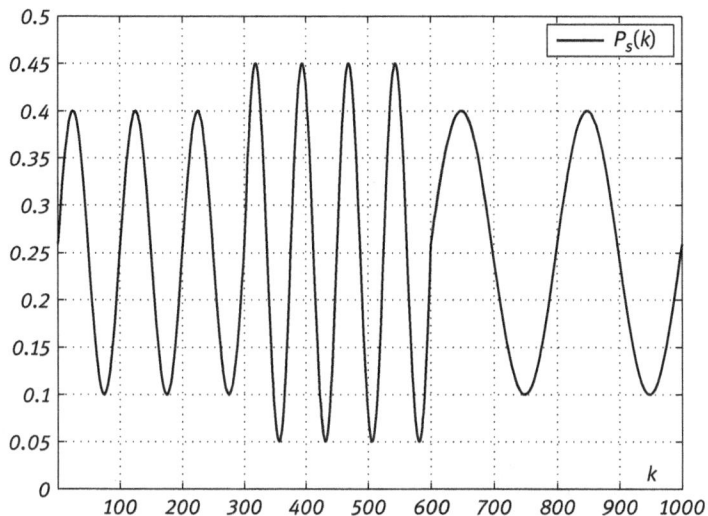

Fig. 2. Evolution of the real non-stationary sinusoidal unknown input.

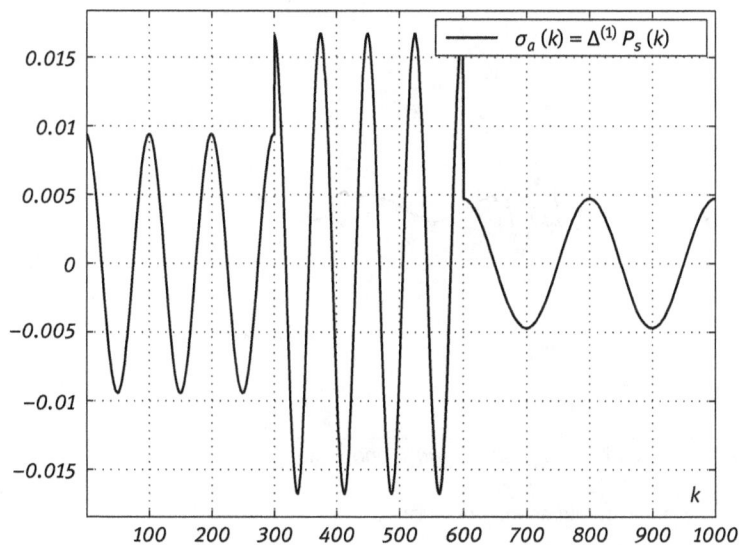

Fig. 3. Evolution of σ_a (k)$=\Delta^{(1)}P_s(k)$.

The estimation errors provided by the proportional Q–integral multiobserver are globally affected by the unknown input. We notice that the generated estimation errors remain globally bounded but not null.

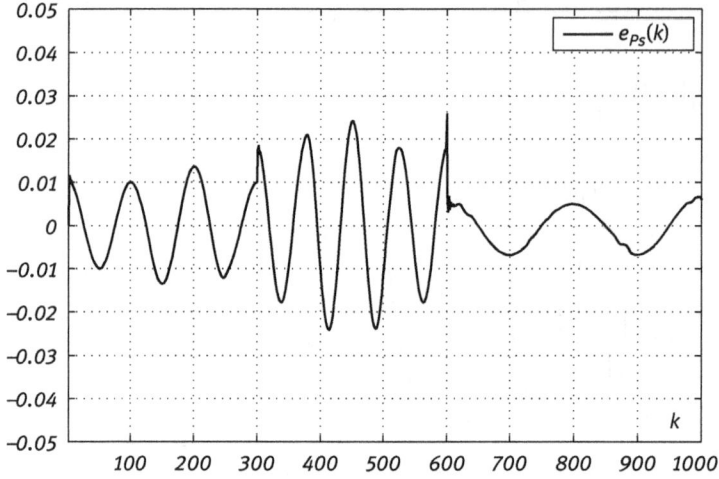

Fig. 4. Evolution of the unknown input estimation error.

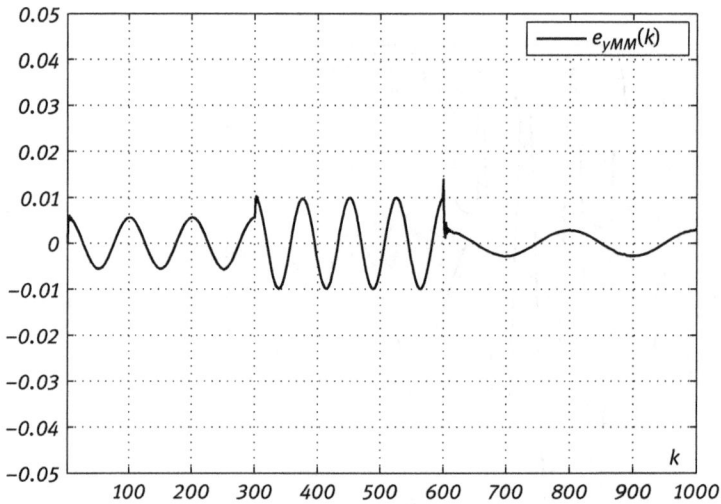

Fig. 5. Evolution of the multimodel output estimation error.

4 Robustness Enhancement in the Case of Non-stationary Sinusoidal Unknown Inputs

In this section, we discuss a particular study that provides the convergence conditions of the estimation error and ensures a good decoupling from non-stationary sinusoidal unknown inputs even in the presence of polynomial unknown inputs with known degree.

To this end, let us consider the sinusoidal signal specified by the above expression:

$$P_s(k) = P_m \sin\left(\frac{2k\pi}{T_s} + \varphi_s\right) \tag{28}$$

where P_m and T_s are respectively the amplitude and the period of the sinusoidal unknown input.

The expression of the qth difference is described via the following equations:

$$\Delta^{(q)} P_s(k) = (-1)^{\left(\frac{q-1}{2}\right)} \left[2\sin\left(\frac{\pi}{T_s}\right)\right]^q P_m \cos\left(\frac{2k\pi}{T_s} + \frac{q\pi}{T_s} + \varphi_s\right),$$

$$\textit{for odd values of } q. \tag{29}$$

$$\Delta^{(q)} P_s(k) = (-1)^{\left(\frac{q}{2}\right)} \left[2\sin\left(\frac{\pi}{T_s}\right)\right]^q P_m \sin\left(\frac{2k\pi}{T_s} + \frac{q\pi}{T_s} + \varphi_s\right),$$

$$\textit{for peer values of } q. \tag{30}$$

Hence, considering equations (29) and (30), we can notice that the q successive differences of $P_s(k)$ drifts the following term:

$$\varsigma = \left[2\sin\left(\frac{\pi}{T_s}\right)\right]^q P_m, \tag{31}$$

The term ς represents the maximum amplitude of $\Delta^{(q)} P_s(k)$. Thereafter, to insure the convergence of $\Delta^{(q)} P_s(k)$ to the neighborhood of zero, we should satisfying some conditions making ς tends to zero. Thus, it is noteworthy that there exists a limit period T_{sl} which checks $\left|2\sin(\frac{\pi}{T_{sl}})\right| < 1$, such as for $T_s > T_{sl}$, a judicious determination of $q \in \mathbb{N}^*$ guaranteed the convergence of ς towards ε_d infinitely small.

For any sinusoidal unknown input $P_{s_j}(k)$ characterized by its amplitude P_{m_j} and its period T_{s_j} which check:

$$P_{m_j} \leq P_m \tag{32}$$

$$T_{s_j} \geq T_s > T_{sl} \tag{33}$$

it becomes easy to deduce that:

$$|\varsigma_j| \leq |\varsigma|$$

where: $\varsigma_j = \left[2\sin(\frac{\pi}{T_{s_j}})\right]^q P_{m_j}$

Thus, if we bring back the value of ς towards zero by determining suitably the value of Q, we can consider that $\Delta^{(Q)}P_s(k)$ and $\Delta^{(Q)}P_{s_j}(k)$ converge towards ε_d infinitely small.

The Q value tending the Q–th difference of a non-stationary sinusoidal unknown inputs to the neighborhood of zero can be determined by the following rule:

$$Q = f_Q \left[\frac{\log(\varepsilon_d) - \log(P_m)}{\log\left(2\sin(\frac{\pi}{T_s})\right)} \right] \tag{34}$$

where f_Q is a function that rounds the elements of r to z. z represents the nearest integers greater than or equal to r.

The function f_Q is defined by:

$$f_Q : \mathbb{R}^* \rightarrow \mathbb{N}^*$$

$$r \rightarrow z$$

Subsequently, for all $T_{s_j} \geq T_s > T_{sl}$ and $P_{m_j} \leq P_m$, the obtained value of Q ensures the convergence of $\Delta^{(Q)}P_s(k)$ towards ε_j infinitely small ($\varepsilon_j \leq \varepsilon_d$).

Remark 1:
The obtained value of Q reflects the number of integral actions ensuring a good decoupling with respect to non-stationary sinusoidal unknown inputs (characterized by its maximum amplitude P_m and its minimum period T_s). This value enables also to decouple perfectly the effect of polynomial unknown inputs with a known degree $d_c = Q\text{-}1$. In the case where the degree of the polynomial unknown inputs d_c is greater than or equal to the number of integral actions obtained via the rule (34), we must consider a Q number of integral actions greater than or equal to $d_c + 1$.

Taking into account the proposed strategy, the estimation errors will be then relatively decoupled from non-stationary sinusoidal unknown inputs and the estimation error (18) becomes as follows:

$$\psi(k+1) = \varphi(k)\,\psi(k) \tag{35}$$

Thus, the objective is to adjust the gain K_r that ensures the convergence of the estimation error to zero and guaranteed the robustness of this error respect to non-stationary sinusoidal unknown inputs.

The convergence conditions are represented in the following theorem [16]:

Theorem 2:
The estimation error (35) between the decoupled state multimodel (5) and the proportional Q–integral multiobservers (8) and (10), tends exponentially to zero, if

there exist two matrices $X = X^T > 0$ and W of appropriate dimensions solutions of the following **LMI's**:

$$\begin{bmatrix} (1-2\alpha)X & (X\Omega - W\theta_i)^T \\ (X\Omega - W\theta_i) & X \end{bmatrix} > 0, \quad i = 1 \ldots N_m \tag{36}$$

If this inequality has a solution for a given rate decay,

$$0 < \alpha < 0.5,$$

the multiobserver gain is determined by:

$$K_r = X^{-1} W \tag{37}$$

4.1 Simulation Results

Two strategies are used in order to decouple the unknown input described by equation (26).

The strategy developed in [16] and [18], consists to decouple the influence of polynomial unknown inputs while attenuating the impact of non-stationary sinusoidal unknown inputs.

However, the proposed strategy is based on the minimization of non-stationary sinusoidal unknown inputs' impact while taking into account the degree of the polynomial unknown input.

Thereafter, by fixing $\varepsilon_d = 10^{-3}$, the number of integral actions which ensures a good decoupling and minimize the effect of the non-stationary sinusoidal unknown inputs on the estimation error can be determined starting from equation (34). Taking into account the characteristics of the non-stationary sinusoidal unknown inputs, the sufficient number of integral actions is equal to Q=3.

Hereafter, in order to show the effectiveness of the proposed multiobserver, there is provided to compare their unknown input and multimodel output estimation errors to those provided by other estimation strategy shown in section (III-C).

Thereafter, Fig. 6 displays the evolution of the 3rd difference of $P_s(k)$:

This figure shows that 3 integral actions make tender $\Delta^{(3)}P_s(k)$ to ϵ_0 lower than ϵ_d. Subsequently, we note that these three integral actions offer a good decoupling of the impact of the unknown input on the estimation errors contrary to the case of consideration of one integral action that presents important values of $\Delta^{(1)}P_s(k)$ as shown in Fig. 3.

The evolutions of the partial models state variables, their estimates and their estimation errors are presents in Figs. 7–11.

These figures show a good estimation because it is clearly seen that the components of the estimated states vectors via the proportional Q–integral multiobserver are coincided with those of the partial models. Otherwise, the estimation errors

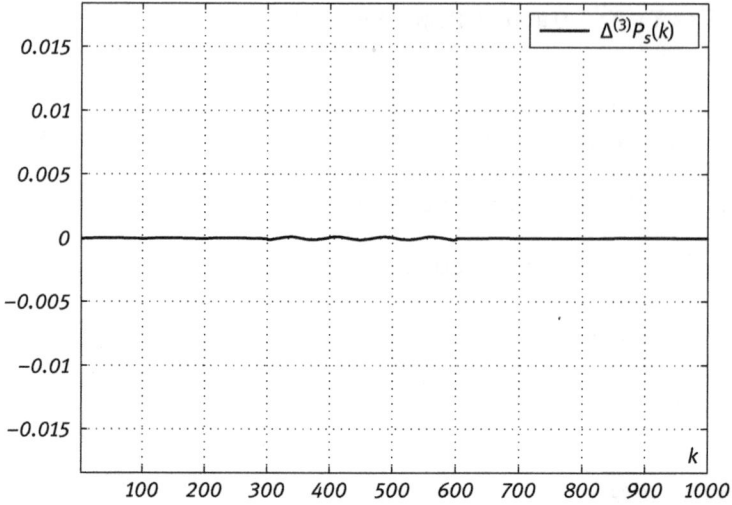

Fig. 6. Evolution of $\Delta^{(3)}P_s(k)$.

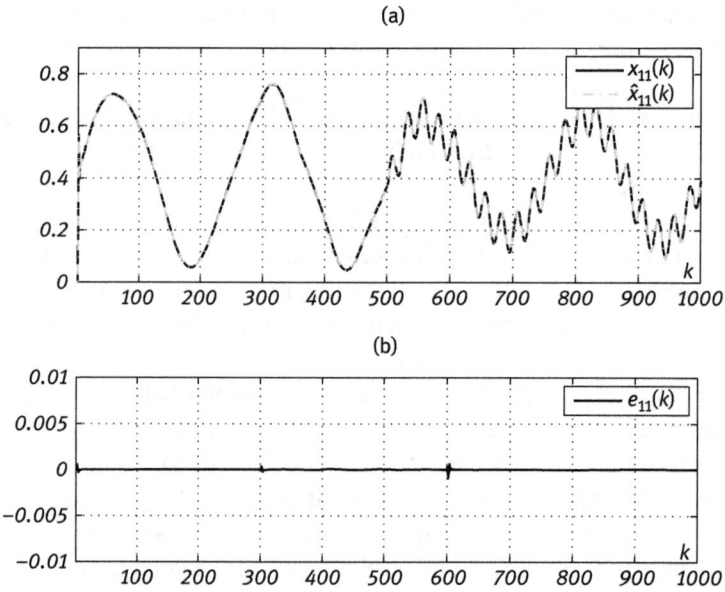

Fig. 7. Partial model 1: (a) Evolutions of the state 1 and its estimate (b) Evolution of the state estimation error.

provided by the proportional Q–integral multiobserver are not globally affected by the unknown input and can be considered negligible.

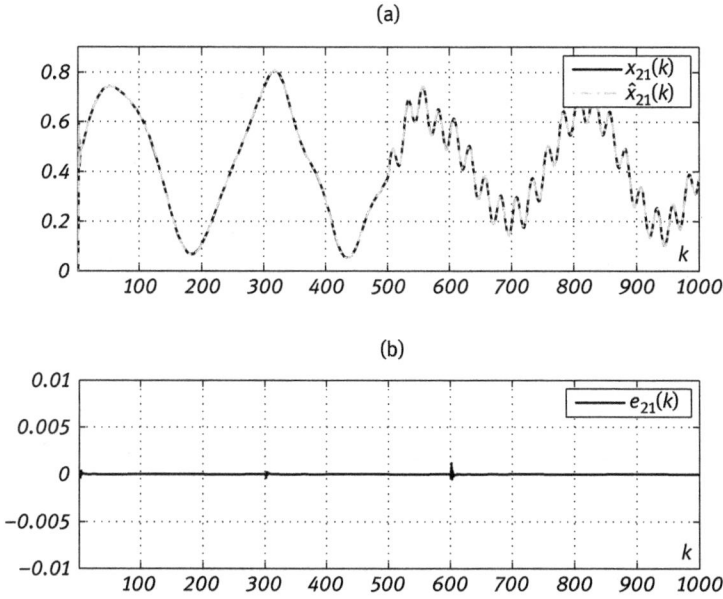

Fig. 8. Partial model 1: (a) Evolutions of the state 2 and its estimate (b) Evolution of the state estimation error.

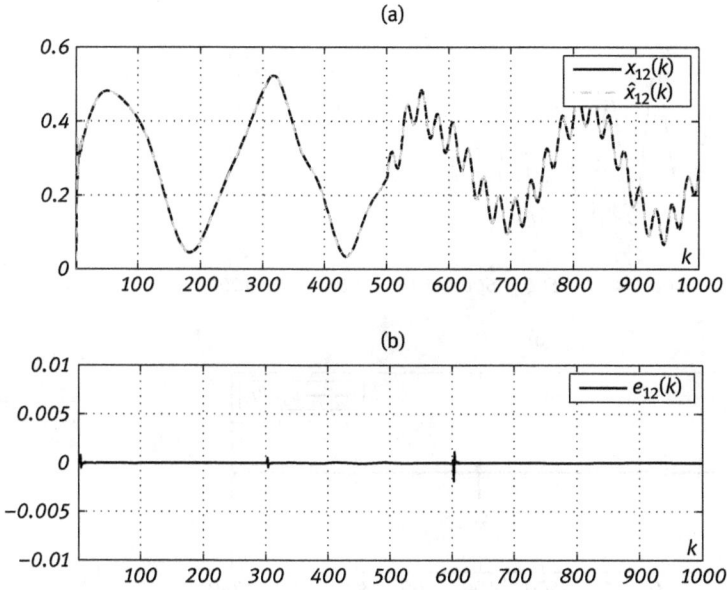

Fig. 9. Partial model 2: (a) Evolutions of the state 1 and its estimate (b) Evolution of the state estimation error.

(a)

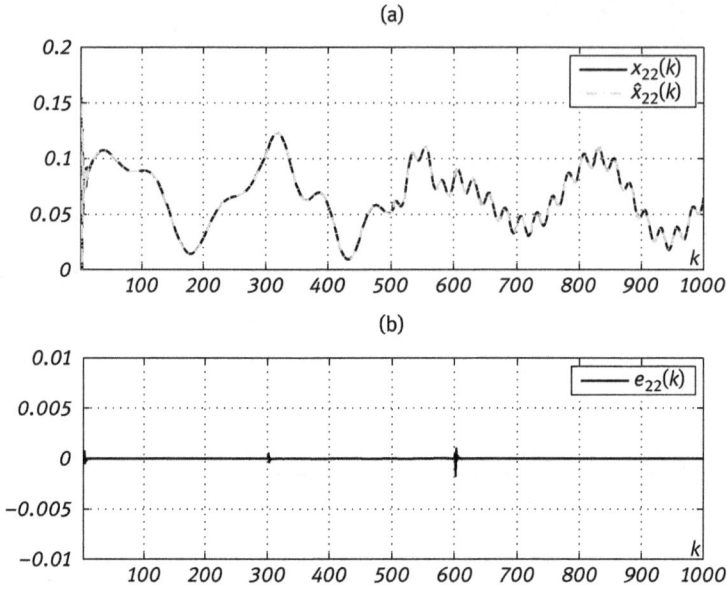

(b)

Fig. 10. Partial model 2: (a) Evolutions of the state 2 and its estimate (b) Evolution of the state estimation error.

(a)

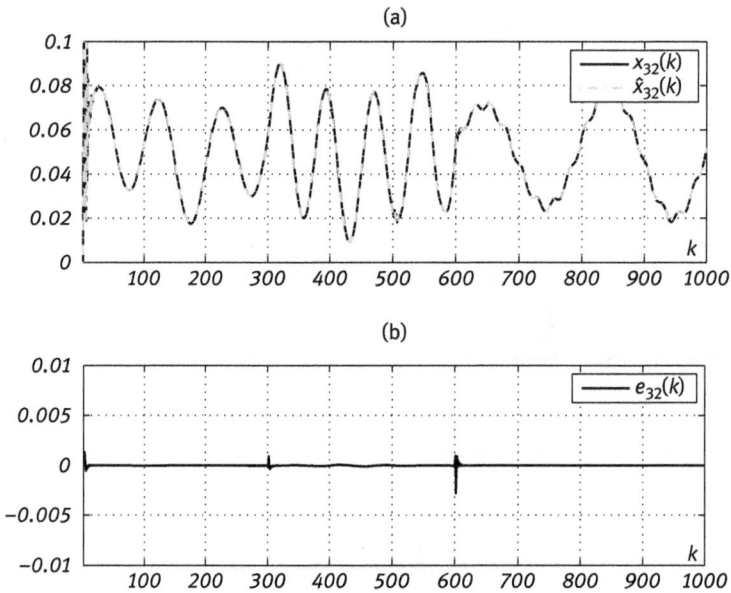

(b)

Fig. 11. Partial model 2: (a) Evolutions of the state 3 and its estimate (b) Evolution of the state estimation error.

(a)

(b)

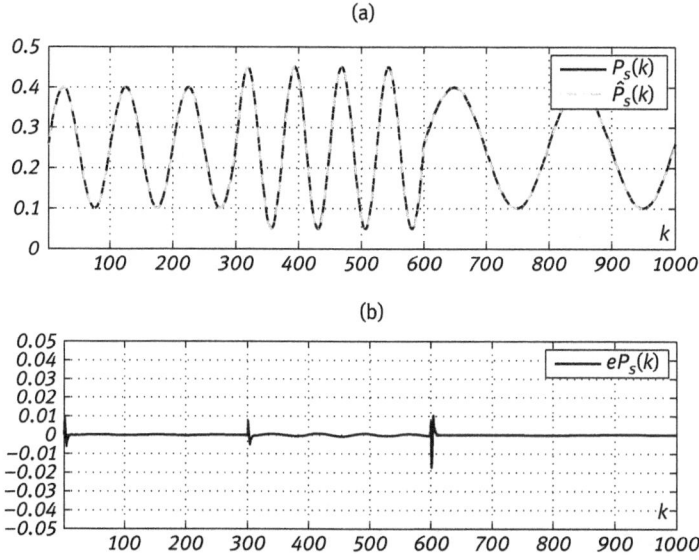

Fig. 12. (a) Evolutions of the non-stationary sinusoidal unknown input and its estimate (b) Evolution of the estimation error.

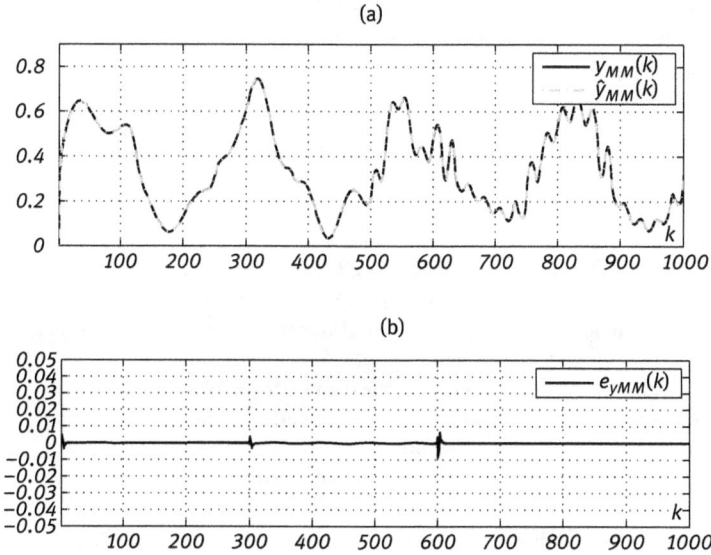

(a)

(b)

Fig. 13. (a) Evolutions of the multimodel output and its estimate (b) Evolution of the output estimation error.

Furthermore, the proposed multiobserver can provides an excellent non-stationary sinusoidal unknown inputs and multimodel output estimation as plotted in Figs. 12 and 13.

These figures let appear a good quality of unknown input and multimodel output estimation and show that the estimation errors remain to the neighborhood of zero. Thus, comparing the estimation errors obtained via the attenuation strategy shown in Figs. 4 and 5, to those generated by the proposed multiobserver, we note that the proposed strategy provides a good estimation errors dynamic with a robust performances.

5 Conclusion

In the present paper, a proportional Q–integral multiobserver is synthesized for the simultaneous non-stationary sinusoidal unknown inputs and state estimation of nonlinear systems described by decoupled state multimodel. The multimodel approach is used to ensure a good compromise between the complexity of the multiobserver synthesis and the accuracy of representation and estimation. The stability and convergence conditions have been formulated using linear matrix inequalities. Finally, a simulation example has provided satisfactory results in terms of convergence of the estimation error and decoupling this estimation error from non-stationary sinusoidal unknown inputs which shows the effectiveness of the proposed robustness enhancement strategy.

Bibliography

[1] M. Allaoui, A. Messaoud, M. Ltaief, and R. Ben Abdennour. Nonstationnary sinusoidal unknown inputs multiobserver for discrete-time nonlinear systems. *International Journal of Sciences and Techniques of Automatic control & computer engineering*, 8(1):1932–1949, 2014.
[2] D. Filev. Fuzzy modeling of complex systems. *International Journal of Approximate Reasoning*, 5(3):281–290, 1991.
[3] Z. Gao, T. Breikin, and H. Wang. Discrete time proportional and integral observer and observer-based controler for systems with both unknown and output disturbances. *Optimal Control Applications and Methods*, 29:171–189, 2008.
[4] S. Ibrir. Robust state estimation with q–integral observers. In *American Control Conference*, 2004.
[5] D. Ichalal, B. Marx, J. Ragot, and D. Maquin. Simultaneous state and unknown inputs estimation with pi and pmi observers for takagi sugeno model with unmeasurable premise variables. In 17th *Mediterranean Conference on Control and Automation, MED*, 2009.
[6] D. Ichalal, B. Marx, J. Ragot, and D. Maquin. Fault tolerant control for takagi-sugeno systems with unmeasurable premise variables by trajectory tracking. In *Proceedings of IEEE International Symposium on Industrial Electronics (ISIE)*, 2010.

[7]   G. P. Jiang, S. P. Wang, and W. Z. Song. Design of observer with integrators for linear systems with unknown input disturbances. *IEEE Electronics Letters Online*, 36(13):1168–1169, 2000.

[8]   D. Koenig. Unknown input proportional multiple-integral observer design for linear descriptor systems: Application to state and fault estimation. *IEEE Transactions on Automatic Control*, 50(2):212–217, 2005.

[9]   J. Korbicz, M. Witczak, and V. Puig. Lmi-based strategies for designing observers and unknown input observers for non-linear discrete-time systems. *Bulletin of the polish academy of sciences technical sciences*, 55(1):31–42, 2007.

[10] Zs. Lendek, J. Lauber, T. M. Guerra, R. Babuska, and B. De Schutter. Adaptive observers for ts fuzzy systems with unknown polynomial inputs. *Fuzzy Sets and Systems*, 161:2043–2065, 2010.

[11] M. Ltaief, A. Messaoud, and R. Ben Abdennour. Optimal systematic determination of models' base for multimodel representation: real time application. *International Journal of Automation and Computing*, 11(6):644–652, 2014.

[12] A. Messaoud. *Sur la représentation et la commande prédictive multimodèles des systèmes complexes*. PhD thesis, Ecole Nationale d'Ingénieurs de Gabès, 2010.

[13] A. Messaoud, M. Ltaief, and R. Ben Abdennour. A new contribution of an uncoupled state multimodel predictive control: Experimental validation on a chemical reactor. *International Review of Automatic Control*, 3(5), 2010.

[14] A. Messaoud, M. Ltaief, and R. Ben Abdennour. Supervision based on a multipredictor for an uncoupled state multimodel predictive control. In *The 6th International conference on electrcal systems and automatic control, JTEA'2010, Hammamet, Tunisia*, 2010.

[15] S. Mondal, G. Chakraborty, and K. Bhattacharyya. Lmi approach to robust unknown input observer design for continuous systems with noise and uncertainties. *International Journal of control, Automation and systems*, 8(2):210–219, 2010.

[16] R. Orjuela. *Contribution à l'estimation d'état et au diagnostic des systèmes représentés par des multimodèles*. PhD thesis, Institue National Polytechnique de Lorraine, France, 2008.

[17] R. Orjuela, B. Marx, J. Ragot, and D. Maquin. Estimating the state and the unknown inputs of nonlinear systems using a multiple model approach. In 16th *Mediterranean Conference on Control and Automation Congress Centre, Ajaccio, France*, 2008.

[18] R. Orjuela, B. Marx, J. Ragot, and D. Maquin. On the simultaneaous state and unknown input estimation of complex systems via a multiple model strategy. *Control theory and applications*, 3(7):877–890, 2008.

[19] T. Takagi and M. Sugeno. Fuzzy identification of systems and its applications to modeling and control. *IEEE Transactions on Systems, Man, and Cybernetics*, 15:116–132, 1985.

[20] A. G. Wu, G. R. Duan, and W. Liu. Proportional multiple-integral observer design for continuous-time descriptor linear systems. *Asian Journal of Control*, 14(2):476–488, 2012.

[21] A. G. Wu, G. Feng, and G. R. Duan. Proportional multiple-integral observer design for discrete-time descriptor linear systems. *International Journal of Systems Science*, 43(8):1492–1503, 2012.

Biographies

Mouhib Allaoui received his Engineering Diploma in Electrical and Automatic Engineering, in June 2012, and the Master degree in Automatic Control and Intelligent Techniques, in December 2012, from National School of Engineers of Gabes-Tunisia. Currently, he is a Ph.D. candidate at Research Unit – CONPRI (Numerical Control of Industrial Processes). His areas of interest deal with multimodel and supervised multicontrol approaches, unknown inputs multiobservers and representation of complex systems.

Anis Messaoud received the Engineering and Master degrees in electrical Engineers and automatic control from National School of Engineers of Gabes-Tunisia, in 2004 and 2006 respectively. In 2010, he obtained his Ph. D. degree in electrical-automatic engineering from the National School of Engineers of Gabes, Tunisia. His specific research interests are in the area predictive control of complex systems, Multimodel and Multicontrol approaches. He is currently an associate professor in the Electric Engineering Department in National School of Engineers of Gabes-Tunisia.

Ridha Ben Abdennour received the Doctorat de Spécialité degree from Higher School of Technical Education in 1987, and the Doctorat d'Etat degree from the National School of Engineers of Tunis-Tunisia, in 1996. He is Professor in Automatic Control at the National School of Engineers of Gabes - Tunisia. He was chairman of the Electrical Engineering Department and the Director of the High Institute of Technological Studies of Gabes. Ridha BEN ABDENNOUR is the Head of the Research Unit of Numerical Control of Industrial Processes and he is the founder and the honorary President of the Tunisian Association of Automatic Control and Digitalization. His research is on Identification, Multimodel & Multicontrol approaches, Numerical Control and Supervision of Industrial Processes. He is the co-author of a book on Identification and Numerical Control of Industrial Processes and he is the author of more than 300 publications. Ridha BEN ABDENNOUR participated in the organization of many Conferences and he was member of some scientific committees of congresses.

M. Bdiwi, J. Suchý, M. Jokesch and A. Winkler

Improved Peg-in-Hole (5-Pin Plug) Task: Intended for Charging Electric Vehicles by Robot System Automatically

Abstract: This paper deals with establishing of the electrical connection between a plug and a receptacle by a robot manipulator for the purpose of charging electrical vehicles. In general, the task of the robot for automatic charging of vehicles consists of two phases. In the first phase, the robot system defines the position of the charging receptacle of the vehicle using vision or infrared system. After that in the next phase, it starts to interact with the environment by connecting the charger plug to the charging receptacle (socket) of the vehicle. However, this phase is not always performed successfully, especially when the socket has complicated shape or consists of multi cores with different sizes. In this paper we will use robot force control to build up the connection. Additionally, an algorithm will proposed which improves the peg-in-hole task by generating spiral motion. The proposed algorithm has shown promising results performed on 5-pin industrial charger plug which is very hard to peg in the socket, even for the human, because it is secure and weatherproof (the plug should cover the whole socket cavity), moreover it has multi cores (5 pins) and it is provided with multiple notches to avoid mismatching between similar pins. In addition to that, the proposed algorithm has assumed that a small vision error could be occurred during estimating the initial position of vehicle's receptacle.

Keywords: Robot manipulator, robot force control, force/torque sensor

1 Introduction

As an alternative resource of energy in the vehicles, using of electricity is increased massively in the new generation of automobiles. According to International Energy Agency, Clean Energy Ministerial, and Electric Vehicles Initiative, there were over 180,000 highway-capable plug-in electric passenger cars and utility vans worldwide until December 2012 [1]. Plug-in electric vehicles means that the batteries which store the electric energy and power the motor are charged by plugging the vehicle socket into an electric power source. Plug-in electric vehicles are classified into two types:

M. Bdiwi, J. Suchý, M. Jokesch and A. Winkler: M. Bdiwi, Fraunhofer Institute for Machine Tools and Forming Technology, Germany, email: mohamad.bdiwi@iwu.fraunhofer.de , J. Suchý, University of Matej Bel, Slovakia, email: jozef.suchy@umb.sk , M. Jokesch, Technische Universität Chemnitz, Germany, email: michael.jokesch@etit.tu-chemnitz.de , A. Winkler, Hochschule Mittweida University of Applied Sciences, Germany, email: alexander.winkler@hs-mittweida.de

De Gruyter Oldenbourg, ASSD – Advances in Systems, Signals and Devices, Volume 5, 2018, pp. 77–88.
DOI 10.1515/9783110470468-005

1. Plug-In hybrid electric vehicles (PHEVs): are powered by an internal combustion engine that can run on conventional or alternative fuel and an electric motor that uses energy stored in a battery. 2. All-Electric vehicles (battery electric vehicles BEVs): use only batteries to store the electric energy that powers the motor. The batteries of both types are requiring recharging by plugging task.

In general, typical EV charging stations require a person who removes the charging adapter from its station and then plugs it in the vehicle's charging socket. As soon as the battery is recharged, he/she should pull out the charger from the socket and return it to the charging station. The charging time could range up to few hours depending on the battery type. Because of the relatively small amount of stored energy in the batteries in comparison with fuel powered vehicles, it should be charged as often as possible when parking, e.g. in public parking garages. Due to these reasons, numerous researches and inventions are performed in order to provide these vehicles with energy effortlessly, automatically and without any human intervention [2], [3] and [4]. All these previous works have proposed an automatic recharging robot system provided with either infrared or vision system for detecting the vehicle and defining the position of the charging receptacle. However, they have considered that charging task will be successfully performed just by connecting the charger plug with the charging socket. In other words, the problem of pegging the plug in the socket has not been deeply discussed.

In this work we propose a robot system for charging vehicles automatically. The robot system plugs the power automatically into the socket of the electric vehicles. This task can be also seen as the peg-in-hole problem which has been often investigated in robotics since many years [5], [6], [7]. In our scenario we assume that finally a camera system identifies the vehicle's position. Going out from this initial target position the plugging task will by executed with the help of robot force control. This paper will focus only on this peg-in-hole task. The proposed algorithm has assumed that a small vision error could be occurred during the estimation of the initial position of vehicle's receptacle. In other words, in the initial position, where the proposed algorithm will start, the notch and the pins of the charger plug are not facing directly the key and the holes of the charging socket. This error should be corrected during the plugging task by robot force control. The proposed algorithm has been applied on plug/socket type IEC 60309 which is very hard for plugging, even for the human.

The paper is organized as follows. In the next section, the experimental setup will be presented. After this, section 3 describes the proposed algorithm the of peg-in-hole task. In section 4 the implementation of the proposed algorithm in a commercial controller and the experimental results will presented. The last section contains discussion and conclusion.

2 Experimental Setup

The experimental setup can be seen in Fig. 1. It consists of the six-axes articulated robot KUKA KR6/2 which is controlled by the industrial robot controller of type KRC2. The robot is equipped with the FT Delta SI-660-60 force sensor by SCHUNK. It is a six component force/torque sensor with the effective measurement range of ±660 N and ±60 Nm for forces and torques, respectively. The sensor is mounted on the manipulator flange and it is protected by a pneumatic robot load limiter. The robot end-effector consists of a two-finger-parallel gripper. With its special fingers it is able to hold the charger plug.

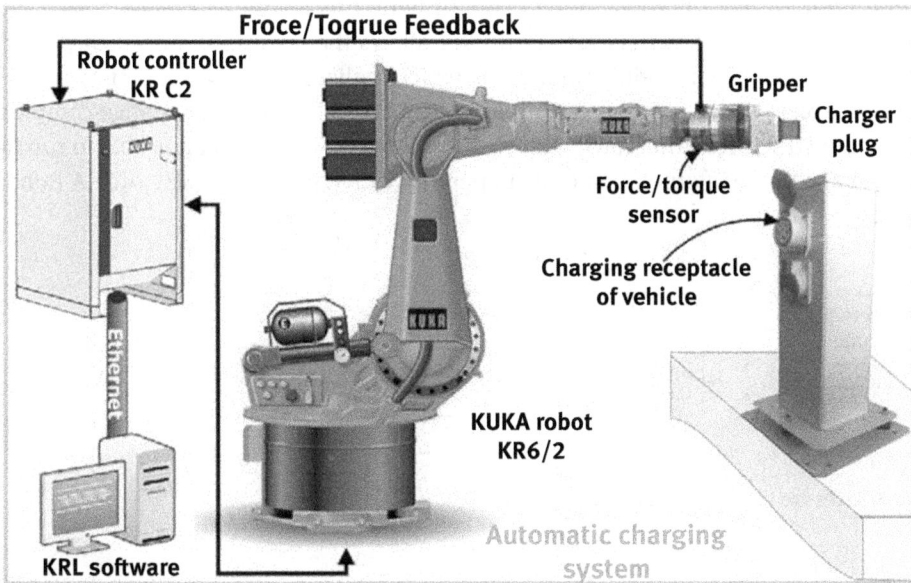

Fig. 1. Experimental setup.

The environment is actually industrial socket mounted on metal board. The implemented environment is simulating the vehicle's receptacle in the real application. The proposed system should peg the charger plug in the socket automatically without any human intervention. As already mentioned, the whole system has used also vision system to define the initial position of the target socket. However, this paper will handle only the problems of peg-in-hole task solved by force control.

Using a commercial robot controller, the implementation of sensor guided robot motion, e.g. force/torque control, will by sometimes difficult, because usually industrial robots are position controlled in joint space using cascade control. The desired positions are provided by the trajectory generator periodically within the interpolation cycle of the robot controller to achieve jerkless point to point or continuous path motions. In the case of continuous path motions like linear or circular interpolation the inverse kinematics of the manipulator have been taken additionally into consideration to calculate the desired joint angels represented by vector q_d to achieve the desired Cartesian position/orientation X_d. For the transient behavior of the closed loop force control it is favorable to have an access point between trajectory generator and the joint position control loops together with the block of inverse kinematics. Then it will be possible to influence the robot motion in force control mode by Cartesian position correction ΔX_d. These correction values act in addition to the values provided by the trajectory generator. This structure of robot force control, which can be seen in Fig. 2, can be also called position based force control [8], [9]. For the robot system chosen here, this feature is available with the KUKA Robot Sensor Interface (RSI) [10]. RSI allows the realization of sensor guided robot motions. It is an additional module which realizes real time signal processing and the access to the position control loops. It is possible to create relative complex controller structures. After having created the whole RSI structure, it is able to run in real time with the interpolation cycle in parallel with the standard KRL (KUKA Robot Language) program.

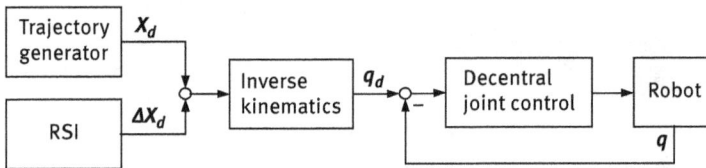

Fig. 2. Access to the position control loops using RSI.

The proposed algorithm has been tested on plug/socket type IEC 60309 which is very hard for plugging even for the human, as shown in Fig. 3. IEC 60309 plug/socket has many safety features which make plugging task almost very difficult. It can be seen in Fig. 3, the IEC 60309 industrial plug/socket is very secure and weatherproof, because the plug should cover the whole socket cavity. Furthermore, pins and slots dimensions are almost the same to permit only proper insertion of plug into socket and the ground pin has a larger diameter than the other pins, preventing the wrong type of plug being inserted in a socket. In addition to that, the plug could not be inserted inside the socket unless the major keyway on the plug aligns with the notch on the socket at the beginning. These entire issues make peg-in-hole task has many challenges. The next section will illustrated how the robot will peg the plug recharger in charging socket automatically even of the previous mentioned difficulties.

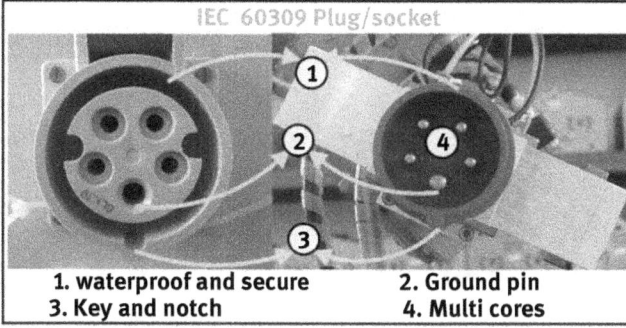

IEC 60309 Plug/socket

| 1. waterproof and secure | 2. Ground pin |
| 3. Key and notch | 4. Multi cores |

Fig. 3. Implemented plug/socket.

3 Proposed Algorithm

3.1 Force Control

Robot force control will be used to build up the connection between plug and socket. In [11] we already investigate different force controllers on the robot system also used in this paper. Although, the proportional controller with positive feedback of the current robot end-effector position has been preferred. With respect of force control error ΔF (input) and Cartesian position correction ΔX_d (output), of course this controller structure has integral behavior. Integral behavior is necessary to contact the environment successfully in the case that the robot end-effector is located in free space. For the purpose of more simplification it may be possible to use also pure integral controllers.

For the present task, force control will be performed only in the translational Cartesian DoFs, which mean that vector of desired forces F_d, vector of current forces F and vector of force control error $\Delta F = F_d - F$ follow to:

$$F_d = \begin{bmatrix} F_{xd} \\ F_{yd} \\ F_{zd} \end{bmatrix}, \quad F = \begin{bmatrix} F_x \\ F_y \\ F_z \end{bmatrix}, \quad \Delta F = \begin{bmatrix} \Delta F_x \\ \Delta F_y \\ \Delta F_z \end{bmatrix}. \tag{1}$$

In this context the vector of controller outputs is:

$$\Delta X_d = \begin{bmatrix} \Delta x_d \\ \Delta y_d \\ \Delta z_d \end{bmatrix} \tag{2}$$

and the integral controllers for x, y and z-direction are represented by controller gains k_x, and k_y and k_z, respectively:

$$\frac{\Delta x(s)}{\Delta F_{xd}(s)} = \frac{k_x}{s}, \quad \frac{\Delta y(s)}{\Delta F_{yd}(s)} = \frac{k_y}{s}, \quad \frac{\Delta z(s)}{\Delta F_{zd}(s)} = \frac{k_z}{s}. \tag{3}$$

3.2 Spiral Motion

Figure 4 illustrates the proposed algorithm for pegging the plug in the socket. As previously explained, the initial position will be defined using vision system. In this phase, we have assumed that the vision results could contain a small error with a maximum of approximately 1 cm. The robot should correct the error automatically with the help of force control.

When the robot has reached the initial position, then the proposed algorithm will be activated and the robot will start to insert the plug-notch in the socket-key. In this phase, the perpendicular axes to the socket will be pure force controlled. The desired value of the contact force is chosen to $F_{zd} = -25\,\text{N}$. The desired values for x and y-direction are set to zero ($F_{xd} = 0$, $F_{yd} = 0$). In these two axes, force control will be additional superimposed by a spiral motion, which is illustrated in Fig. 5. This connection between force and position control may be understood as the approach of parallel position/force control introduces in [12].

During the spiral motion, the system tests if the notch of the plug is inserted in the key of the socket. In the case that the current end-effector position in z-direction differs more than 1 cm from the z-coordinate of the first environment contact, we assumed that the notch is inserted into the key. Otherwise, if the notch is outside of the key, the diameter of the spiral motion should be increased. The maximum diameter of the largest circle in the spiral path in x and y-direction is 1 cm. In other words, the maximum vision error which could be corrected by the proposed algorithm is 1 cm. Hence, when the notch is inserted in the key, the robot controller has to resave the values of the initial position of x and y in order to correct the vision error. The new initial coordinates will be the centric position of the spiral motion in the following subtask.

In the next phase, the proposed algorithm activates force control and spiral motion again. Now the robot has the task to insert the cores of the plug in the holes of the socket. Just like in the previous phase, the perpendicular axis to the socket will be pure force controlled and x and y axes are shared position/force controlled. After every full cycle of the spiral motion, when its diameter reaches the predefined maximum, the covered distance of the end-effector in z-direction as a result of force control has to be analyzed. If this distance is small, e.g. smaller than 2 mm, the desired force value F_{zd} will be increased in steps of 25 N. For the purpose of saving time, the value of

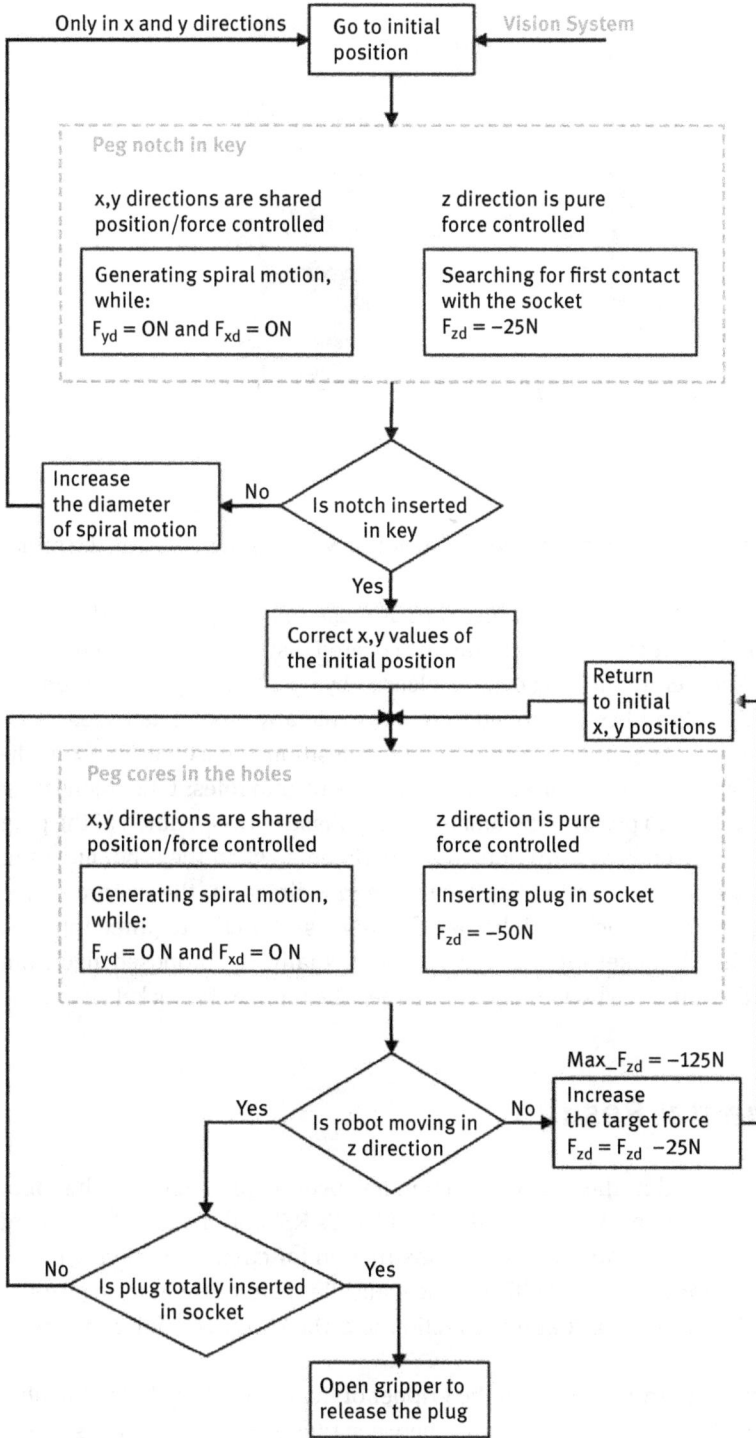

Fig. 4. Peg-in-hole algorithm for connecting plug and socket.

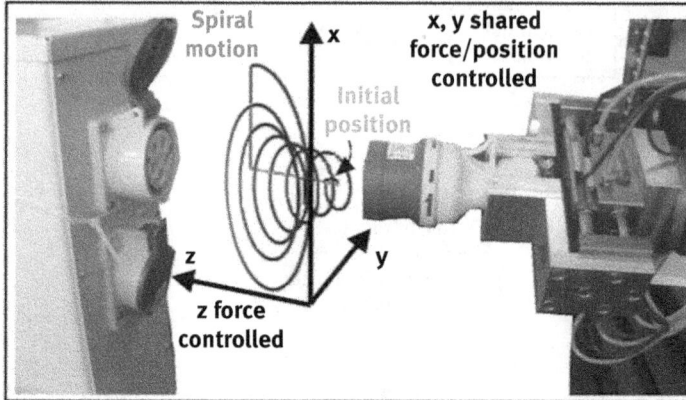

Fig. 5. Spiral motion.

$F_{zd} = 50\,\text{N}$ will be skipped. After this, the spiral motion will start again using the initial diameter.

The peg-in-hole task is finished when the cores are inserted completely in the holes. This situation can be detected by comparing the covered distance of the plug with the length of the socket. Before the plug can be released by open the gripper, the contact force in all direction should be controlled to zero to avoid any kind of damage.

In conclusion, the proposed algorithms (shared position/force control in x, y direction and pure force control in z direction) will play two main roles: 1. Compensating the vision error in order to perform the first inserting phase between notch of the plug and the key of the socket. 2. According to our experiments, when the robot moves in spiral path in x, y directions during applying force in z-direction, the required force for pegging the plug in the socket will be significantly less than the required force for pegging the plug in the socket without that motion. Actually, the proposed algorithm is inspired from the human when he/she tries to peg the plug in the socket.

4 Experimental Results

The algorithm presented in the previous section has been implemented in the robot controller. Force control has been applied using KUKA's RSI and the spiral motion of the end-effector has been realized using the instruction for circular motion which is provided by KUKA Robot Language (KRL). The controller gain of the force controller are chosen to $0.01\,\text{mm}\,\text{N}^{-1}$ for x and y-direction and $0.075\,\text{mm}\,\text{N}^{-1}$ for z-direction, respectively.

At the beginning of the experiment the gripper of the robot which holds the plug is located in front of the socket. The distance between plug and socket with respect to z-direction is approximately 1 cm. Besides, for the simulation of the error of the initial position, caused by the vision system, the plug and socket faces not exactly.

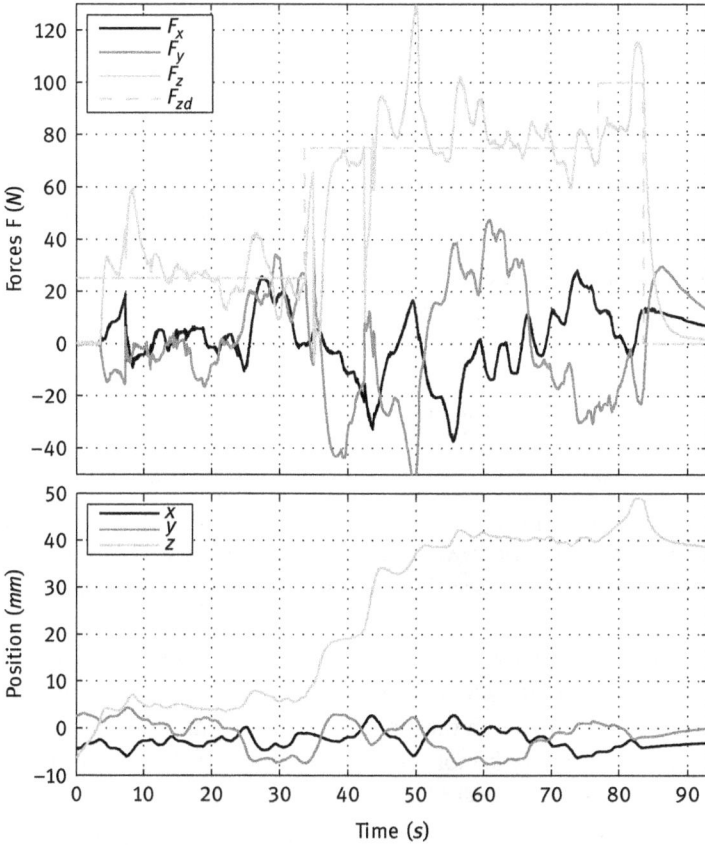

Fig. 6. Forces and position during the peg-in-hole task.

Figure 6 shows the forces acting on the robot end-effector and its position during the peg-in-hole task. In the first phase the desired force has been set to 25 N. After approximately 5 s the first contact between plug and socket has been found. In x and y-direction the spiral motion and zero force control work in parallel and the notch of the plug is inserted into the key of the socket. After approximately 35 s we increase the desired force to 75 N. Regarding the z-position of the end-effector, it can be seen that the plug is inserted step by step in the socket which is a result of joint force control and spiral motion. In the present experiment it is necessary to increase the desired force value once gain up to 100 N to insert to socket completely into the plug. This can be recognized on the end-effector motion from 40 mm to 50 mm (z-coordinate) at the time around 80 s.

Before the plug can be released by open the gripper, it may be suitable to activate zero force control in all three translational directions. With it some tensions will be reduced and damages will be avoided.

5 Conclusion

This paper presents an approach to build up the connection between a plug and a socket automatically by a robot manipulator. In the future it should be used for charging electrical vehicles e.g. in public parking garages.

We have been assumed that the approximately position of the socket at the car can be detected by a vision system. Our approach deals only with the insertion of the plug into the socket. For this purpose we prefer the combination of force control and a spiral motion of the plug, which can be understood as parallel or shared force/position control.

For the verification of the proposed algorithm a test bed was configured, which consist of an articulated robot and a pillar which contains an industrial electric socket. In the experiment we show that the robot which holds the corresponding industrial plug is able to insert the plug into the socket successfully. However, there are some further efforts:

So we could determine in the experiments that the proposed algorithm is something sensitive with respect to the parameters of the spiral motion e.g. the velocity and the acceleration of the circular parts. From this point of view it seems to be necessary to investigate the parameterization of the motion commands and the force controllers with the aim to make the peg-in-hole process faster and more robust.

Bibliography

[1]   iea (International Energy Agency, Clean Energy Ministerial, and Electric Vehicles Initiative) *Global EV Outlook 2013 - Understanding the Electric Vehicle Landscape to 2020*. April, 2013.

[2]   S. Hollar and E. Hollar.   *United States patent No. 7,999,506B1, System to Automatically Recharge Vehicles with Batteries*. 2011.

[3]   K. Cornish. *United States patent No. 2012/0233062 A1, Method and Process for an Electric Vehicle Charging System to Automatically Engage the Charging Apparatus of an Electric Vehicle*. 2012.

[4]   P. Joshué, F. Nashashibi, B. Lefaudeux, P. Resende and E. Pollard. Autonomous Docking Based on Infrared System for Electric Vehicle Charging in Urban Areas. *Sensors*, 13:2645–2663, 2013.

[5]   S. N. Simunovic. Parts mating theory for robot assembly. In *Proc. of the 9th International Symposium on Industrial Robots*, 183–193, 1979.

[6] I. Godler, Y. Takahashi, K. Wada and R. Katoh. Peg-and-hole task by robot with force sensor: Simulation and experiment. In *Proc. of the International Conference on Industrial Electronics, Control and Instrumentation*, 2:980–985, 1991.
[7] S. Jörg, J. Langwald, J. Stelter, G. Hirzinger and C. Natale. Flexible Robot-Assembly using a Multi-Sensory Approach. In *Proc. of the IEEE International Conference on Robotics and Automation*, 4:3687–3694, 2000.
[8] J. De Schutter and H. Van Brussels. Compliant Robot Motion II. A Control Approach Based on External Control Loops. *International Journal of Robotics Research*, 7(4):18–33, 1988.
[9] E. Dégoulange and P. Dauchez. External Force Control of an Industrial PUMA 560 Robot. *Journal of Robotic Systems*, 11(6):523–540,1994.
[10] KUKA Roboter GmbH. *KUKA.RobotSensorInterface (RSI) 2.1*, 2007.
[11] A. Winkler and J. Suchý. Position Feedback in Force Control of Industrial Manipulators – An Experimental Comparison with Basic Algorithms. In *Proc. of IEEE International Symposium on Robotic and Sensors Environments*, 31–36, 2012.
[12] S. Chiaverini and LSciavicco. The Parallel Approach to Force/Position Control of Robotic Manipulators. *IEEE Transactions on Robotics and Automation*, 9(4):361–373, 1993.

Acknowledgment: The investigations presented in this paper were funded by the Federal Ministry of Education and Research of Germany under the project number 03IPT505A. The responsibility with respect to the content of the publication has the authors.

Biographies

Mohamad Bdiwi received the Bachelor degree in Control and Automation Engineering from Aleppo University in Syria, 2007. He received his Ph.D. degree in Robotic Systems Department at Chemnitz Technical University, Chemnitz, Germany, in 2014. The research field of Dr. Bdiwi is oriented toward human robot interaction, visual servoing and force control. He works since 3 years as project manager in Fraunhofer Institute IWU in Chemnitz, Germany.

Jozef Suchý received his Ing. and PhD. Degrees in electrical engineering from Slovak University of Technology in Bratislava, Slovakia in 1973 and 1978, respectively. Until 1996 he was with the Institute of Control Theory and Robotics, Slovak Academy of Science, Bratislava. Until April 2015 he was professor and head of the Department of Robotic Systems, Faculty of Electrical Engineering and Information Technology, University of Technology Chemnitz, Germany. Since April 2015 he is with the Department of Computer Science, Faculty of Natural Science, University of Matej Bel, Banská Bystrica, Slovakia.

Michael Jokesch received the Bachelor of Science (B.Sc.) degree in electrical engineering and the Master of Science (M.Sc.) degree in energy and automation systems from Technische Universität Chemnitz, Germany, in 2011and 2013, respectively. Currently he works as a researcher at the department of Robotics and Human-Machine-Interaction, Faculty of Electrical Engineering and Information Technology, Technische Universität Chemnitz.

Alexander Winkler received the Dipl.-Ing., Ph.D. and State doctorate degrees in electrical engineering / automation from Technische Universität Chemnitz, Germany, in 2000, 2006 and 2015, respectively. Until 2014, he was with the Department of Robotic Systems, Faculty of Electrical Engineering and Information Technology, Technische Universität Chemnitz, Germany. He is currently a Professor with the Department of Automation, Hochschule Mittweida, University of Applied Sciences, Mittweida, Germany.

L. Sellami, S. Zidi and K. Abderrahim

Active Mode Estimation via Clustering Algorithm for Switched Linear Systems

Abstract: The work presented in this paper deals with the active mode identification problem for switched linear systems based on a measurement data set. This problem is an issue closely related to the classification problem of input–output measure data. Indeed, each data group associated with their most appropriate sub-model presents a mode (discrete state) of operation. Therefore, we propose a method for discrete state estimation based on a clustering algorithm combined with a decision mechanism. The clustering algorithm provides the class centers which will be exploited by the decision mechanism in order to identify the discrete state. Simulation results are presented to illustrate the performance of the proposed method.

Keywords: Switched linear systems, System Identification, Mode estimation, Clustering algorithm.

1 Introduction

In several innovative technologies, the interaction increasingly important between physical processes (evolution models involving continuous signals) and digital processes (computers, software, logic components, etc.) led in automatic, to the appearance and the construction of the so-called hybrid systems [1, 2]. Hybrid systems are defined as dynamic systems do explicitly and simultaneously interact nature phenomena both continuous and event (discret). The mode knowledge representing the discrete dynamic evolution of the hybrid dynamic systems at any instant is a decisively information which enables a simplified application of the various results coming from the fields of identification, control, stability analysis, and state estimation. This paper addresses the problem of discrete state estimation of switched linear systems a class of hybrid systems whose continuous dynamics is defined by a discrete-time state model. There are two methods for identification. The first technique, called a offline estimation method, consist in identify the discrete state over a observation horizon N of data. Thus, the continuous state $x(k)$ estimation, modeling continuous dynamic, is based on the technique developed in [10]. However, the second technique is online and simultaneous estimate of discrete state and continuous state such as in [3].

L. Sellami, S. Zidi and K. Abderrahim: L. Sellami, National Engineering school of Gabes, University of Gabes, Tunisia, email: sellami_lamaa@yahoo.fr , S. Zidi, College of Business and Economics, Qassim University, KSA, email: S.Zidi@qu.edu.sa , K. Abderrahim, National Engineering school of Gabes, University of Gabes, Tunisia, email: kamelabderrahim@yahoo.fr

De Gruyter Oldenbourg, ASSD – Advances in Systems, Signals and Devices, Volume 5, 2018, pp. 89–104.
DOI 10.1515/9783110470468-006

Numerous methods have been proposed to solve this problem. The approach presented in [4], for linear switched system a class of SDH, is based on an application of the Multivariable Output-Error State Space (MOESP) identification technique [5] for the estimation of sub-models, in a noisy environment. In this approach, the authors assume that the discrete state and the number of sub-models are known a priori. Further, the switching instants are separated by a minimum dwell time. In [6], the considered model does not suffer of state databases matching problem such as a switched linear systems model. Indeed, the different linear sub-models do not interact but they evolve independently of each other so that one can represent each of them in an arbitrary base. But the application of the presented method for the identification of this model (according to the authors formulation) may require a very important dwell time. Thus, there are many work in different search domain such as works of identification [4, 6], control [15, 16], stability analysis [10, 12–14] and state estimation [8, 9], make the knowledge of the switching mechanism as a basic assumption.

However, other researchers have opted recently to other approaches based on artificial intelligence, learning and classification. These new methods showed quite encouraging results for several automatic issues. Boukharouba [7] proposed a modeling approach of hybrid systems based on the classification. In this work, the switching mechanism is modeled by a piecewise linear functions defining the validation boundaries of each sub-models. The parameters adaptation of these functions is provided by a new incremental and decremental classification algorithm multi-class support vector. Ackerson and Fu [18] are the first which considered the determination problem of the active mode through state estimation in noisy environment. A recent result for Markov jump linear system (MJLS) a class of hybrid system, developed in [17], is presented to the state estimation in the presence of polyhedral bounded disturbances. The approaches [19–21] are proposed to identify the switching instants for a class of switched linear systems described in the state form. In these works, a identification recursive algorithm and dynamic classification algorithm are used to estimate the discrete state.

In this paper, a offline estimation method of activate mode for the discrete-time linear switched systems is proposed from a finite data set. The proposed method is based on two stages. First, a clustering technique selects the representative points of different classes (submodels). Second, this points will be used by a decision mechanism in ordre to identify the mode (discrete state). Indeed, the discrete state is determined by the minimum value of s criteria which involves the difference between the system output and the center of the appropriate class.

The paper has the following structure. Section 2 deals with the problem formulation. The active mode estimation is developed in section 3, and the continuous dynamic estimation is presented in section 4. In section 5,we present the simulations in order to illustrated the effectiveness of the proposed method.

2 Problem Formulation

We consider a class of hybrid dynamics systems described by the following state space model:

$$\begin{cases} x(k+1) = A_i x(k) + B_i u(k) \\ y(k) = C_i x(k) \end{cases} \tag{1}$$

where $i \in S = \{1, \ldots, s\}$ is the discrete state of the system, s is the number of submodels, $x(k) \in \mathbf{R}^n$, $y(k) \in \mathbf{R}$ and $u(k) \in \mathbf{R}$ are respectively the continuous state, the output and the input of the system, A_i, B_i and C_i are the parameter matrices associated with the submodel indexed by i.

Given data $\{x(k), y(k)\}_{k=1}^N$ generated by a switched linear model of the form as in (1), we are interested in determining a comportement model of switched linear system. To deal with this problem, we make the following assumptions:
(i)   No model has been specified for the switching mechanism. In fact, the switches can be exogenous, deterministic, state-driven, event-driven, time-driven or totally random.
(ii) The order n is the same for all submodels and unknown.
(iii) The number of submodels s is unkown.

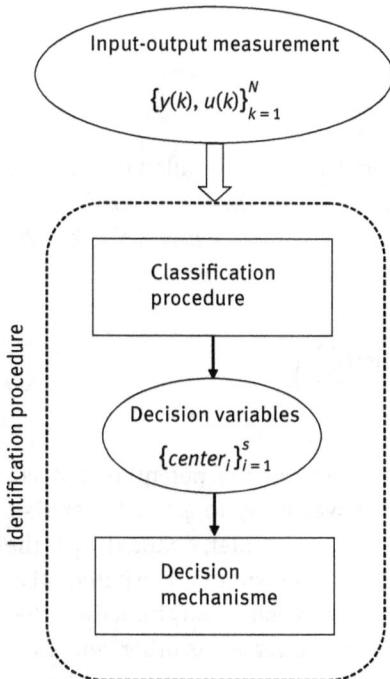

Fig. 1. Structure of the proposed estimation method.

According to Fig. 1, the active mode estimation based on a classification procedure and a decision mechanism. The classification procedure provides the class centers $(\varphi_1^*, \varphi_2^*, \ldots, \varphi_s^*)$ which will be exploited by the decision mechanism in order to identify the discrete state $\hat{q}(k)$. However, to estimate the continuous dynamic, we use the switched observer approach. The stability analysis of the switched linear observer involves a Poly-Quadratic stability principle [11]. This concept allows to check asymptotic stability of the state vector estimation $\hat{x}(k)$ by mean of polytopic quadratic Lyapunov functions.

3 Active Mode Estimation

Most of existing methods consist in estimating the active discrete state using a set of data defined on an observation horizon. This set is assumed to be pure i.e the elements of this set are generated by the same submodel. However, this assumption is far from being achieved in practice because these elements may be generated by several submodels. To overcome this problem, they introduce the concept of dwell time to ensure that the data belong to the same submodel.

In this paper, we propose a novel approach which consists in associating the current output $y(k)$ to its appropriate submodel. This assignment is insured thanks to a clustering algorithm.

3.1 Cluster Centers Selection

We suggest the use of the Chiu's clustering algorithm [17, 19], also called the method of mountains [20, 21]. In this work, we extended this method to classify a set of regression's vectors $\varphi(k) \big(\varphi(k) = [y(k)\ y(k-1)\ u(k-1)]^T, k = 1, \ldots, N \big)$ [28, 29] which consists in associating to each $\varphi(k)$ a potential P_k defined by:

$$P_k = \sum_{\substack{l=1 \\ l \neq k}}^{N} exp\left(\frac{-4\|\varphi(k) - \varphi(l)\|^2}{r_a^2} \right). \tag{2}$$

where r_a is a positive coefficient that control the decay of the potential. Indeed, the potential decreases exponentially when $\varphi(l)$ moves away to $\varphi(k)$. Otherwise, the coefficient r_a defines the radius of a class. The first class center, denoted φ_1^* is the data whose potential P_1^*, expressed by equation (5), is the maximum. The potential of the neighborhood points of the center while gradually decreasing away from the latter. To avoid selecting the data in the neighborhood of the center φ_1^* as other centers of classes, the classification procedure changes the value of each potential given by the following formula:

$$P_k \leftarrow P_{k-1} - P_1^* exp\left(\frac{-4\|\varphi(k) - \varphi_1^*\|^2}{r_b^2}\right). \tag{3}$$

The parameter r_b ($r_b > 0$) must be strictly greater than r_a to promote the operation on the selection of other distinctly different classes of first and its close. The center of second class is selected as the data having the maximum modified potential given by relation (6). Let φ_2^* the second center and P_2^* the associate modified potential.

Similarly, we selected the cth center φ_c^* having P_c^* as a potential and the potentials are modified by the following formula:

$$P_k \leftarrow P_{k-1} - P_c^* exp\left(\frac{-4\|\varphi(k) - \varphi_c^*\|^2}{r_b^2}\right). \tag{4}$$

Chiu introduced two positive parameters ε_1 and ε_2 ($\varepsilon_1 > \varepsilon_2$) to condition the choice of centers. Indeed, the decision of the selection procedure of the various centers of classes obeyed, at each step, the following inequalities:

- If $P_c^* > \varepsilon_1 P_1^* \Rightarrow$ selection authorized.
- If $P_c^* < \varepsilon_2 P_1^* \Rightarrow$ selection achieved.
- If $\varepsilon_2 P_1^* \leq P_c^* \leq \varepsilon_1 P_1^*$ and if:

$$\frac{Min\left(|\varphi_c^* - \varphi_1^*|, |\varphi_c^* - \varphi_2^*|, \ldots, |\varphi_c^* - \varphi_{c-1}^*|\right)}{r_a} \leq 1 - \frac{P_c^*}{P_1^*}$$

where φ_c^* is the current center and $\varphi_1^*, \varphi_2^*, \ldots, \varphi_{c-1}^*$ are already the selected centers. The center to retain corresponds, in these conditions, to the maximum value of potentials after rejection of the current value P_c^* (the value of P_c^* was set to zero).

3.2 Decision Mechanism

Once the centers φ_c^* of classes are selected, we proceed in estimation of discrete state such as the index of the submodel that generated the pair of data $(u(k), y(k))$ in the sense of a certain decision criterion $\mathcal{J}_i(k)$ designed by:

$$\hat{q}(k) = \underset{i=1,\ldots,s}{arg\ min}\mathcal{J}_i(k). \tag{5}$$

The criterion $\mathcal{J}_i(k)$ can be chosen in many different ways. Since no model has been specified for the discrete state dynamics in switching linear system (1). To assignment of the elements $\varphi(k)$ to appropriate class, we proceed in a simple calculation of distances d_{ki} between each point $\varphi(k)$ of these centers φ_i^* for $i = 1, \ldots, s$.

$$d_{ki} = \|\varphi(k) - \varphi_i^*\|_2 \quad \text{for } i = 1, \ldots, s \tag{6}$$

$\varphi(k)$ belongs to the class whose the center corresponds to the distance minimum $d_{kc} = \min_{i=1,\ldots,s} (d_{ki})$. In practice, the normalized distance d_{ki_n} given by the relation (6) are considered.

$$d_{ki_n} = \frac{d_{ki}}{\|\varphi_{max} - \varphi_{min}\|_2} \quad \text{for } i = 1, \ldots, s \tag{7}$$

φ_{max} and φ_{min} are respectively the maximum and minimum values of the data $\varphi(k)$. Thus, $\mathfrak{J}_i(k)$ is can be expressed here as a normalized distance, given by equation (7).

Before summarizing the discrete state estimation algorithm, we must take the following assumption.

Assumption 1 At each instant index k and for any couple (i, j) of discrete states indexes,

$$\mathfrak{J}_i(k) = \mathfrak{J}_j(k) \Rightarrow i = j. \tag{8}$$

The objective of assumption 1 is obviously to remove any ambiguity in the inference of the discrete state if the true submodel are exactly known.

In summary, the proposed approach consists of two steps: one is an offline procedure which allows to determine the class centers using the Chui's algorithm and two is an online operations which consists in estimating the discrete state based on the exploitation of the class centers via a decision mechanism.

---

algorithm 1 : Discret state estimation

Offline part:
Inputs: data $\{y(k), u(k)\}_{k=1}^N$
Calculate the potential P_k as in (2) for $k = 1, \ldots, N$
Select the first center y_1^*
repeat
    Calculate the modified potential as in (3)
    Select the cth center
until $(\varepsilon_2 P_1^* \le P_c^* \le \varepsilon_1 P_1^*)$ and $\left(\frac{Min(|y_c^* - y_1^*|, |y_c^* - y_2^*|, \ldots, |y_c^* - y_{c-1}^*|)}{r_a} \le 1 - \frac{P_c^*}{P_1^*} \right)$
Online part:
for $k = 1, \ldots, N$ **do**
    Compute criteria $\mathfrak{J}_i(k)$ as in (3) for $i = 1, \ldots, s$
    Estimate the discrete state $\hat{q}(k)$ such as $\mathfrak{J}_{\hat{q}(k)}(k) = \min_{i=1,\ldots,s} \mathfrak{J}_i(k)$
end for

---

4 Continuous Dynamic Estimation

The continuous behavior of hybrid dynamic system, considered in this paper, is modeled by a switched observer whose the dynamic is expressed as follow:

$$\begin{cases} \hat{x}(k+1) = A_i\hat{x}(k) + Bu(k) + L_i(y(k) - \hat{y}(k)) \\ \hat{y}(k) = C_i\hat{x}(k) \end{cases}. \tag{9}$$

The gain matrices L_i, $i = 1, \ldots, s$, have to be computed such that the estimation state $\hat{x}(k)$ is asymptotically converged to the system state $x(k)$ whatever the initial conditions, i.e:

$$\forall \varepsilon(0) \in R^n \lim_{k \to \infty} \varepsilon(k) = 0. \tag{10}$$

where $\varepsilon(k) = x(k) - \hat{x}(k)$ is the estimation error and it's dynamic behaviors defined by:

$$\varepsilon(k+1) = (A_i - L_iC_i)\varepsilon(k). \tag{11}$$

The switched observer design reduces to the computation of the gain matrices L_i, $i \in \{1, \ldots, s\}$, ensuring the asymptotic stability for the switched system (10). To solve this problem, we use the concept of Poly-Quadratic stability [11]. This reduces to check stability by mean of particular quadratic Lyapunov functions tacking into account the switching nature of system (1). To recall this concept, the estimation error dynamic becomes:

$$\varepsilon(k+1) = \tilde{A}(\xi_k)\varepsilon(k). \tag{12}$$

The structure of dynamical matrix $\tilde{A}$ is assumed to depend in polytopic way on the parameter ξ_k:

$$\tilde{A}(\xi_k) = \sum_{i=1}^{s} \xi_k^i \tilde{A}_i. \tag{13}$$

where

$$\tilde{A}_i = A_i - L_iC_i$$

and the components of the parameter vector ξ_k appear as indicator functions given by [9]:

$$\xi_k^i = \begin{cases} 1 & \text{when the switched system} \\ & \text{is described by matrix } A_i \\ 0 & \text{otherwise} \end{cases} \tag{14}$$

with $i \in \{1, \ldots, s\}$ and

$$\xi_k = \left[\xi_k^1, \ldots, \xi_k^s\right]$$

To check asymptotic stability of system (11), Poly-quadratic stability uses Lyapunov function with a polytopic structure similar to that of the system description:

$$V(\varepsilon(k), \xi_k) = \varepsilon^T(k) P(\xi_k) \varepsilon(k), \quad \text{with } P(\xi_k) = \sum_{i=1}^{s} \xi_k^i P_i \tag{15}$$

where P_i, $i = 1, \ldots, s$, are symmetric positive definite constant matrices of appropriate dimensions. Using this fact, the following theorem gives sufficient condition to built such a switched observer.

Theorem [9] If there exist symmetric matrices S_i, matrices F_i and G_i solutions of:

$$\begin{pmatrix} G_i + G'_i - S_j & G_i'A_i - F_i'C_i \\ A_i'G_i - C_i'F_i & S_i \end{pmatrix} > 0 \quad \forall(i,j) \in \{1, \ldots, s\}^2 \tag{16}$$

then a switched observer (8) for system (1) exists and the resulting gains L_i are given by $L_i = G_i'^{-1}F_i'$ with $P_i = S_i^{-1}$.

5 Numerical Examples

We now present two simulation examples to illustrate the performance of the proposed method.

5.1 Example 1

In this first example, a switched linear system presented is composed of two-mode where the dynamics are governed by the following equations [14]:

$$A_1 = \begin{bmatrix} 0.5 & 1 \\ 0 & 0.5 \end{bmatrix}, \quad B_1 = \begin{bmatrix} 0 \\ 1 \end{bmatrix}$$
$$C_1 = \begin{bmatrix} 1 & 0 \end{bmatrix} \quad \text{if} \quad u(k) < 0$$

$$A_2 = \begin{bmatrix} 0.7 & 1 \\ -0.5 & 0.5 \end{bmatrix}, \quad B_2 = \begin{bmatrix} 0 \\ 1 \end{bmatrix} \tag{17}$$
$$C_2 = \begin{bmatrix} 1 & 0 \end{bmatrix} \quad \text{if} \quad u(k) \geq 0$$

The control signal considered, in this simulation, is given in the Fig. 2. It is easy to remak that the evolution of the switched signal is randomly.

For camputing the gains of the continuous observer, the LMI (16) is feasible, we obtain:

$$L_1 = \begin{bmatrix} -0.0942 \\ 0.1413 \end{bmatrix}, \quad L_2 = \begin{bmatrix} 0.0982 \\ -0.2201 \end{bmatrix}$$

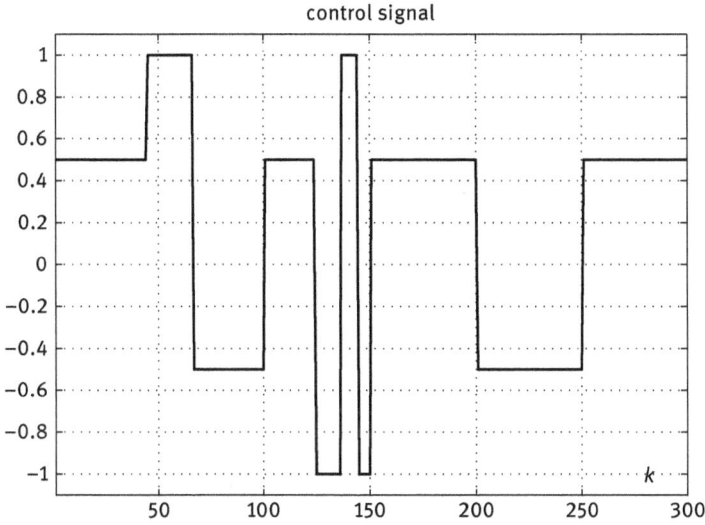

Fig. 2. The switched signal evolution.

The state and the estimated state evolutions are depicted in Fig. 4 which shows the estimated states tracks the true states. Figure 3 illustrates the estimation error evolutions. We remark that these errors converges to zeros which proves the performance of the proposed method.

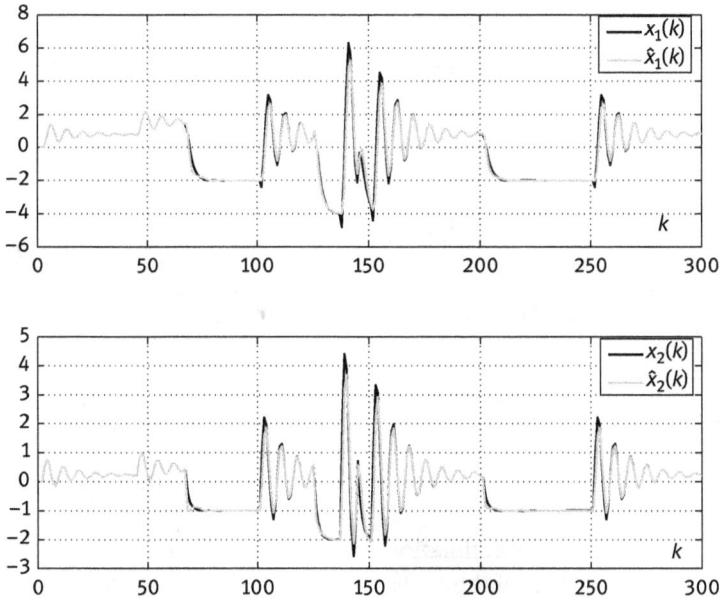

Fig. 3. The state vector evolution and its estimated.

the evolution of e1

the evolution of e2

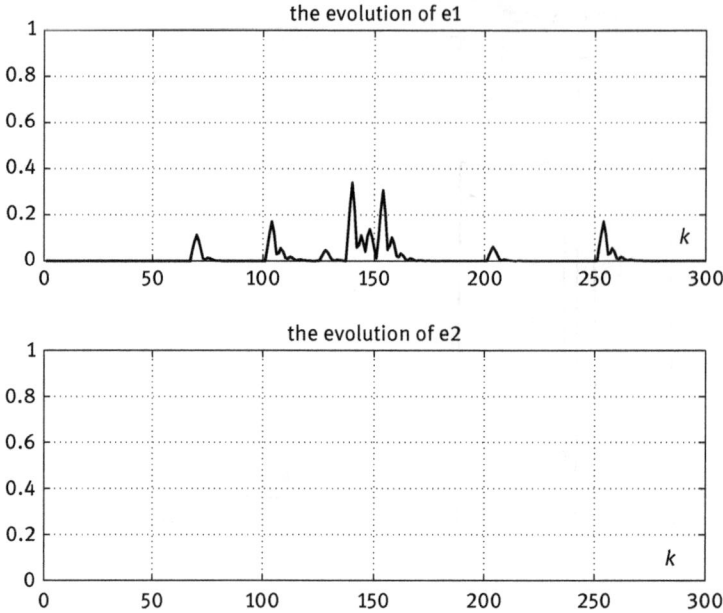

Fig. 4. The state vector's estimation errors.

5.2 Example 2

This second example illustrates the state estimation of a system having three states given by [8]. The parameter matrices are given as follow:

$$
A_1 = \begin{bmatrix} -0.84 & 0 & 0.27 \\ 0.38 & -0.33 & 0.07 \\ -0.1 & 0.55 & 0.44 \end{bmatrix}, \quad B_1 = \begin{bmatrix} 0.11 \\ 0.28 \\ 0.52 \end{bmatrix}
$$

$$
C_1 = \begin{bmatrix} 0.62 & 0.77 & 0.82 \end{bmatrix}, \quad \text{if } x_3(k) \geq 0
$$

$$
A_2 = \begin{bmatrix} -0.39 & 0.07 & -0.13 \\ 0.90 & -0.39 & -0.41 \\ 0.51 & -0.32 & 0.59 \end{bmatrix}, \quad B_2 = \begin{bmatrix} 0.25 \\ 0.14 \\ 0.92 \end{bmatrix}
$$

$$
C_2 = \begin{bmatrix} 0.19 & 0.62 & 0.51 \end{bmatrix} \quad \text{if } x_3(k) < 0
$$

$$(18)$$

In this example, we applied the input sequence to be the realization of a random vector process Fig. 2.

The behavior of the discrete state estimation and the real mode evolution, are presented in Fig. 5, prove the viability of the proposed method to identify the discrete state.

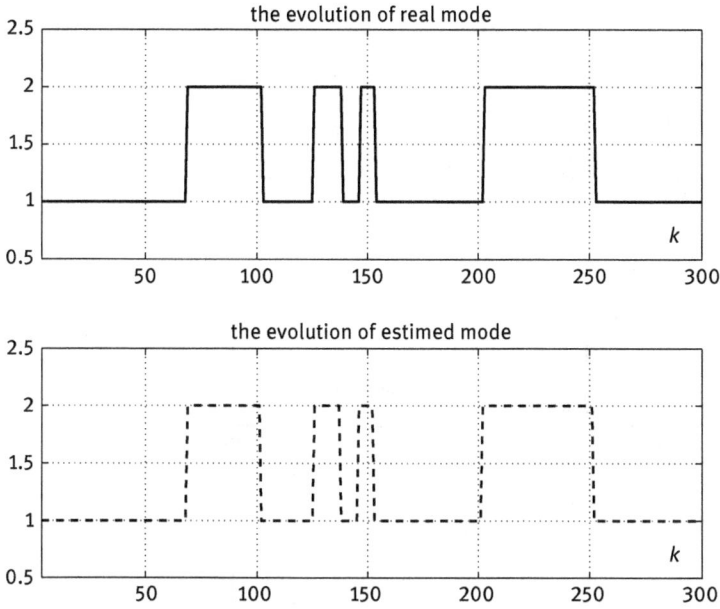

Fig. 5. Upper plot: The real mode. Lower plot: The estimation mode.

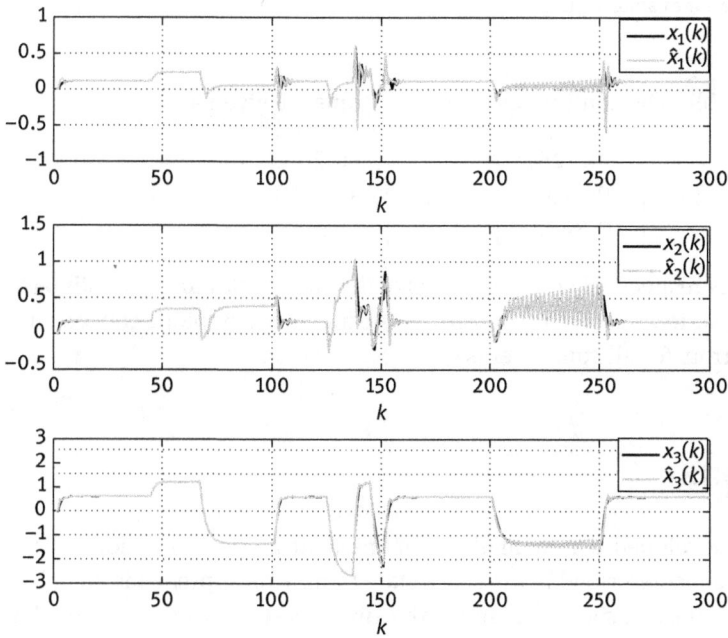

Fig. 6. The evolution of state vector's components and their estimated.

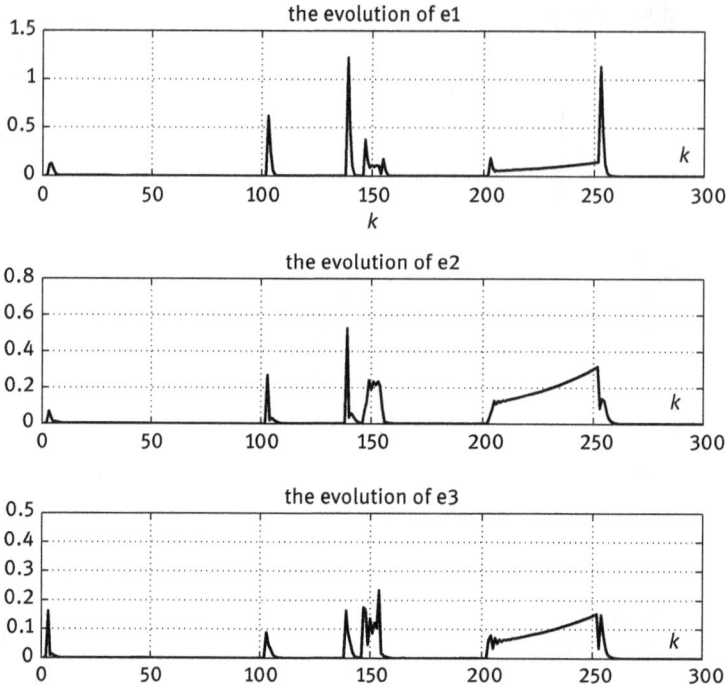

Fig. 7. The estimation error of state vector.

The LMI (16) is feasible. The continuous observer's gains are given as follow:

$$L_1 = \begin{bmatrix} -0.5231 \\ 0.1735 \\ 0.3472 \end{bmatrix}, \quad L_2 = \begin{bmatrix} -0.2045 \\ 0.1750 \\ 0.5465 \end{bmatrix}$$

Figures 6 and 7 gives respectively the evolutions of the state vector and the estimated state vector and the state vector estimation errors. We find the asymptotic convergence of the estimation errors for all components of the state vector.

6 Conclusions

In this paper, we developed a new method for the actif mode estimation for the switched linear systems described in the space state form. This approach is an offline estimation method, it's consists in identify the discrete state over a observation horizon N from a finite data set based on two stages. First, we exploit a Chiu clustering algorithm to determine the number of classes and these centers. Then, the class centers will be explored by a decision mechanism in ordre to identify the mode (discrete state). In fact, the discrete state is determined by the minimum value of s criteria which involves the difference between the system output and the center of the appropriate class. Simulation results are presented to illustrate the performance of the proposed method.

Bibliography

[1]   M. Egerstedt and B.M (Eds), *Hybrid systems*, Computation and Control (HSCC) Proceedings of HSCC 2008, Springer Verlag, St. Louis MO, USA, 2008.

[2]   B. De Schutter and W. Heemels and A. Bemporad, *Equivalence of hybrid dynamical models*, Automatica, vol(37), pp 1085–1091, 2001.

[3]   L. Sellami and K. Abderrahim, *Classification and statistical learning for detecting of switching time for switched linear systems*, Journal of Innovation in Digital Ecosystems, vol (2), pp 13–19, 2015.

[4]   V. Verdult and M. Verhaegen, *Subspace identification of piecewise linear systems*, In Conference on Decision and Control, Atlantis, Paradise Island, Bahamas, 2004.

[5]   M. Verhaegen, *Subspacemodel identification part 3: Analysis of the ordinary output-error statespace model identification algorithm.* International Journal of Control, 58, pp 555–586, 1993.

[6]   K. M. Pekpe, G. Mourot, K. Gasso, and J. Ragot, *Identification of switching systems using change detection technique in the subspace framework*, In Conference on Decision and Control, Atlantis, Paradise Island, Bahamas, 2004.

[7]   K. Boukharouba, *Modélisation et classification de comportements dynamiques des systèmes hybrides*, Ph.D., University of Science and Technology of Lille1, 2011.

[8]   A. Alessandri and P. Coletta, D*esign of observers for switched discrete time linear systems*, in Proceedings of the 2003 American Control Conference, Denver, CO, USA, pp 2785–2790, 2003.

[9]   F. Bejarano, A. Pisano, and E. Usai, *Finite-time converging jump observer for switched linear systems with unknown inputs*, Nonlinear Analysis: Hybrid Systems, vol. 5, no. 2, pp 174–188, 2011.

[10] J. Daafouz, G. Millerioux and C.A Iung, *poly-quadratic stability based approach for linear switched systems*, Special Issue of Mathematics and Computers in Simulation 58, march 2002, pp 295–307.

[11] L. Daafouz and J. Bernussou, *Parameter dependent lyapunov functions for discrete time systems with time varying parametric uncertainties* Journal of Systems and Control Letters 43/5,pp 355–359, August 2001.

[12] A.A. Agrachev and D. Liberzon, *Lie-algebraic stability criteria for switched systems*, SIAM J. Control Optim., vol. 40, No. 1, pp 253–269, 2001.

[13] U. Boscain, *Stability of planar switched systems: the linear single, input case*, SIAM J. Control Optim., vol. 41, No. 1, pp 89–112, 2002.

[14] M.S. Branicky, *Multiple Lyapunov functions and other analysis tools for switched and hybrid systems*, IEEE Trans. Automat. Control, vol. 43, No. 4, pp 475–482, 1998.

[15] Z. Sun, S. Ge and T. Lee, *Controllability and reachability criteria for switched linear control*, Automatica, vol. 38, pp 775–786, 2002.

[16] G. Xie, D. Zheng and L. Wang, *Controllability of switched linear systems*, IEEE Trans. Automat. Control, vol. 47, No. 8, pp 1401- 1405, 2002.

[17] Hao Wua, Wei Wanga, Hao Yea and Zidong Wanga, *State estimation for Markovian Jump Linear Systems with bounded disturbances*, In Automatica 49, pp 3292–3303, 2013.

[18] G. A. Ackerson and K. S. Fu, *On state estimation in switching environnements*, IEEE Transactions on Automatic Control, vol. 15, No. 1, pp 10–17, 1970.

[19] D. Mincarelli, T. Floquet and L. Belkoura, *Active mode and switching time estimation for switched linear systems*, 50th IEEE Conference on Decision and Control and European Control Conference (CDC-ECC) Orlando, FL, USA, 2011.

[20] R.V. Lopes, G.A. Borges and J.Y. Ishihara, *New Algorithm for Identification of Discrete-Time Switched Linear Systems*, American Control Conference (ACC) Washington, DC, USA, 2013.

[21] Y. Tian, T. Floquet, L. Belkoura and W. Perruquetti, *Switching time estimation for linear switched systems: an algebraic approach*, 48th IEEE Conference on Decision and Control and 28th Chinese Control Conference, ShangHai, China, 2009.

[22] L. Daafouz and J. Bernussou, *Parameter dependent lyapunov functions for discrete time systems with time varying parametric uncertainties* Journal of Systems and Control Letters,pp 355–359, August 2001.

[23] F. Rosenqvist and A. Karlström, *Realisation and estimation of piecewise-linear output-error models*, Automatica 41, pp. 545–551, 2005.

[24] S.L. Chiu, *Fuzzy model identification based on cluster estimation*, Journal of intelligent and fuzzy systems 2, pp. 267–278, 1994.

[25] A. Shigeo, *Neural networks and fuzzy systems theory and applications*, Kluwer Academic Publishers, 1997.

[26] R.R. Yager and D.P. Filev, *Approximate clustering via the mountain method*, IEEE Transactions on Systems, Man and Cybernetics 24, no. 8, pp 1279–1284, 1994.

[27] R.R. Yager and D.P. Filev, *Generation of fuzzy rules by mountain clustering*, Journal of intellegent and fuzzy systems 28, pp 209–219, 1994.

[28] M. Ltaief, K. Abderrahim, R. Ben Abdennour and M. Ksouri, *Contributions to the multimodel approach: systematic of a model's base and valities estimation*, Journal of Automation and System Engineering 2, 2008.

[29] M. Ltaief, A. Messaoud and R. Ben Abdennour, *Optimal Systematic Determination of Models'Base for Multimodel Representation: Real Time Application*, International Journal of Automation and Computing, vol 11(6), pp 644–652, 2014.

Biographies

Lamaa Sellami received her engineer degree in electrical-automatic engineering and the M. Eng. degree in automatic and smart techniques from the National School of Engineers of Gabes (ENIG), Tunisia in 2005 and 2007, respectively. Currently, she is a Ph. D. candidate at Research Unit of Numerical Control of Industrial Processes (CONPRI), University of Gabes, Tunisia. Her research interests include hybrid system and identification.

Salah Zidi has got his Master in new technologies of dedicated IT systems in 2004 from National Engineering School of Sfax (ENIS), (Tunisia) and his PhD in information systems and computer science in 2007 from University of Sciences and Technologies of Lille (France). He has been a Head of Research & Development at Archimed Group in France between 2010 and 2014. Before that, he has been an assistant professor of computer science at University Lille 1– France since 2007. He has joint college of business and economics at Qassim University since 2014. He published papers in peer-reviewed journals and has several papers published in international journals and many research projects in progress.His research interests cover topics like Artificial intelligence, classification and machine learning, Support Kernels Regression, feature extraction, OCR, decision support systems, transport reconfiguration, scheduling.

Kamel Abderrahim received the B. Eng. degree in electrical engineering from the National School of Engineers of Gabes (ENIG), Tunisia in 1992, and the M. Eng. degree in automatic control from Higher School of Sciences and Techniques of Tunis, Tunisia (ESSTT) in 1995, and the Ph. D. degree in electrical engineering from National School of Engineers of Tunis, Tunisia (ENIT) in 2000, and the Habilitation in electrical engineering from the University of Gabes in 2009. He has been a member of Laboratory of Numerical Control of Industrial Processes (LACONPRI) at the ENIG since 1995. He joined the ENIG as an assistant professor in 2000, and now he works as a professor at the ENIG. From 2002 to 2005, he was the director of the Electrical Engineering Department at the ENIG. And from 2005 to 2011, he was the director of Higher Institute of Industrial Systems, Tunisia (ISSIG). His research interests include nonlinear process modeling, identification, and control.

R. Zaier

Humanoid Locomotion Control and Reflex Using Van der Pol and Piecewise Linear Oscillators

Abstract: Design of a network of oscillators to control the locomotion of a legged robot is a complex problem dealing with inverse kinematics and dynamics of more than 20 degree of freedom system. An attempt to simplify the design of a locomotion controller is introduced by combining a Piecewise Linear Oscillator (PWLO) and the Van Der Pol Oscillator (VDPO). The VDPO is considered as a master oscillator as it generates the rolling motion pattern. The pitching and swing motions are generated by the PWLO. The VDPO parameters are selected such that the oscillator has a stable limit cycle. By introducing a small perturbation parameter into the VDPO equation, the convergence rate to the stable limit cycle can be controlled. In addition, the locomotion controller is designed as a neural network coupled to the robot dynamics through sensory system to exhibit natural looking motion. To enhance better robustness of the network, proportional derivative controllers are added to the system. Furthermore, the reflex against large disturbance that cannot be compensated by a conventional feedback controller during locomotion is considered.

Keywords: Humanoid, Van der Pol oscillator, piecewise linear oscillator, central pattern generator CPG, limit cycle.

1 Introduction

Locomotion of a legged robot can be thought as a complex problem difficult to solve with too many parameters hard to tune as well. Indeed, conventional neural networks, which are structurally built by connectionists, suffer less flexibility in implementation and in tuning the neurons' connections weights. A network of neural oscillators that consists of master oscillator coupled to several oscillators is considered in this paper. This fascinating idea in dealing with locomotion problem has already been approached by some researchers [1–4]. Although significant progress has been made using small numbers of tightly coupled neurons, those approaches have some limitations in implementing and adapting them to a changing environment. For instance, in quadruped or humanoid robots, joint control signals are more complex than the oscillator based models can generate. Moreover, the

R. Zaier: Mechanical and Industrial Engineering Department, Sultan Qaboos University, Sultanate of Oman, email: zaier@squ.edu.om

De Gruyter Oldenbourg, ASSD – Advances in Systems, Signals and Devices, Volume 5, 2018, pp. 105–124.
DOI 10.1515/9783110470468-007

stability of a walking robot requires controlling many joints, and thus, too many parameters of the coupled oscillators need to be tuned [5]. Some trials towards simplification of locomotion pattern generator have been conducted [6, 7] using piecewise linear oscillators. The research framework in [7] is based on piecewise linear functions that shape the locomotion pattern using recurrent neural network and sensory feedback. Although the method provides a comprehensive knowledge about the robot dynamics and allows easy implementation of sensory feedback and much flexibility to include reflexes capability to the humanoid robot, the locomotion controller is too structural that needs much neurons and wires to build. The work in [6] deals with the phase response properties of Matsuoka oscillator, which has been interpreted as a piecewise linear oscillator with phase resetting control.

In this paper we will extend the research result in [6] by introducing the Van der Pol oscillator [8] abbreviated here as "VDPO" to the neural network of the locomotion controller. The VDPO is considered as the central pattern Generator "CPG" of the rolling motion pattern [2], which will be referred to as the master oscillator. The parameters of the VDPO will be smoothly modulated to enable the robot to exhibit smooth change of behavior such as to switch from walking to standing. The network to be built is based on both the "integrate and fire" neuron model [9] and the recurrent neural network language defined in [10]. The work is expected to provide a pretty compact network of oscillators, simple to tune, while generating both rolling and pitching motion profile. In addition, the neural network can be coupled to the robot dynamics through sensory system to exhibit natural looking motion. Typical design of humanoid robot includes sole sensor, gyro sensor, and accelerometer [11]. Proportional derivative (PD) controllers made of two neurons with input from Zero Moment Point (ZMP)and Gyro sensors will be added to the network to improve the robustness of the locomotion controller against perturbation caused by terrain irregularity.Furthermore,in this paper the reflex against large disturbance during locomotion is considered, which cannot be compensated by a conventional feedback. This reflex consists of generating a particular motion triggered by gyro sensor signal.

The rest of this paper is organized as follows. Section 2 presents the rolling motion pattern generator using PWLO and VDPO. Also it describes the robot dynamics coupled to oscillators; Sections 3 presents the motion generator for pitching and swing motion. Section 4 presents the implementation of the reflex against large disturbance. Section 5 shows the experimental results, and section 6 is conclusion.

2 Rolling Motion Pattern Generation

To begin with, we investigate the stability of the robot simplified as an inverted pendulum when standing at one leg that can be expressed by:

$$\frac{d^2\theta}{dt^2} + 2\zeta\frac{d\theta}{dt} - \mu\theta = u(\theta) \tag{1}$$

where $\theta \in \mathbb{R}$ is the counterclockwise angle of the inverted pendulum to the vertical, $\zeta \in \mathbb{R}$ is damping ratio, $\mu = \frac{g}{l} \in \mathbb{R}$, and $u \in \mathbb{R}$ is the control input to the system. Let $\theta = \alpha - \alpha_s$, where α is the angular position of the hip joint and α_s is its value when the projected center of mass (COM) is inside the support polygon. Then, equation (1) can be rewritten as follows;

$$\frac{d^2\alpha}{dt^2} + 2\zeta\frac{d\alpha}{dt} - \mu\alpha = u(\alpha) \tag{2}$$

2.1 Primitive Motion

Continuous Piecewise Linear (PWL) functions have been proven to be a powerful tool in modeling and analyzing nonlinear systems. For instance, to fit the dynamics of the robot described by (2) for small oscillations, the rolling motion of a humanoid robot can be approximated by a sine-wave. But, in general, the trajectories for the biped joints are not sine functions with only one frequency component. For such situations, a combination of more primitive functions is more interesting. Equation (3) represents the first order model of a neuron. Using this model it is straightforward to express any higher order ODE and its coupling with other ODE as detailed in Zaier and Nagashima 2002.

$$\varepsilon_i\frac{da(t)}{dt} + a(t) = c(t) \tag{3}$$

where $c(t)$ is the input signal, $a(t)$ is the activation function. The input signal $c(t)$ can be expressed by (4) as time series of N piecewise-linear functions:

$$c(t) = \sum_{i=1}^{N} c_i(t - t_i) \tag{4}$$

2.2 PWLO Generating Rolling Motion Pattern

For the sake of reducing the computing cost and ease the implementation of the controller, the rhythmic motion can be then generated with regards to the rolling motion as in [6], which has a trapezoidal form and smoothed using a delay as in (3).

$$u(t_i, \omega_i) = \omega_i(t - t_i)\left[u_s(t - t_i) - u_s\left(t - t_i - \frac{1}{\omega_i}\right)\right] + u_s\left(t - t_i - \frac{1}{\omega_i}\right) \tag{5}$$

where $u_s(\cdot)$ is the unit step function, $\omega_i > 0$ represents the slope of the function $u(.)$ between t_i and $\left(ti + \frac{1}{\omega_i}\right)$. The approximate solution $\alpha(t)$ can be formulated, therefore, as a function of time delay ε, joint angular velocity ω, walking period T, and the rolling amplitude α_s.

$$\alpha = f(\varepsilon, \omega, T, \alpha_s) \tag{6}$$

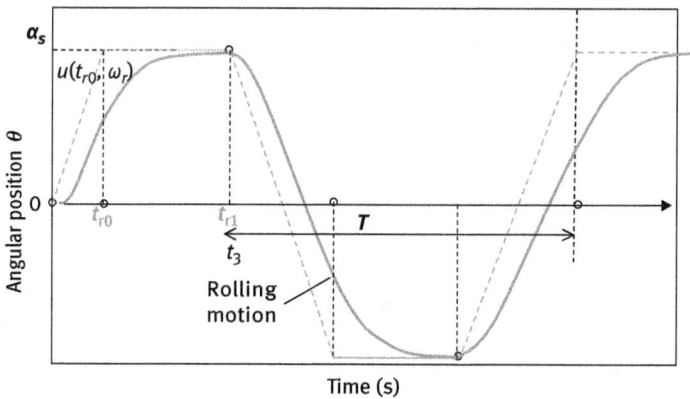

Fig. 1. The pattern of rolling motion [6].

Therefore, the rolling motion can be expressed by (7) and shown in Fig. 1.

$$\varepsilon\frac{d\alpha(t)}{dt} + \alpha(t) = \alpha_s[u(t_{ro}, 2\omega_r) - 2u(t_{ro} + (n + f_1)T, \omega_r)$$
$$+ 2u(t_{ro} + (n + f_2)T, \omega_r) - u(t_f, 2\omega_r)] \tag{7}$$

where t_{ro} and t_f are the times at the start and the end of the rolling motion, respectively. n is the number of walking steps. f_1 and f_2 are relative times with respect to the gait. The significance of (7) is in its real-time implementation using the Recurrent Neural Network (RNN) language developed in [10, 12]. This RNN language is suitable for the programmers to reflect the biological process in generating robot motion. In contrast to the mathematical notations used in [1, 3], this proposed language can express the learning process of a motion, just by changing connections, and their weights for a given RNN circuit. The neural network is assumed to solve problems using 4 types of operations namely, summation, multiplication by a constant, introduction of a time delay constant, and switching. Using this language, several RNN circuits can be easily designed termed Central Pattern Generators (CPG) as detailed by [10].

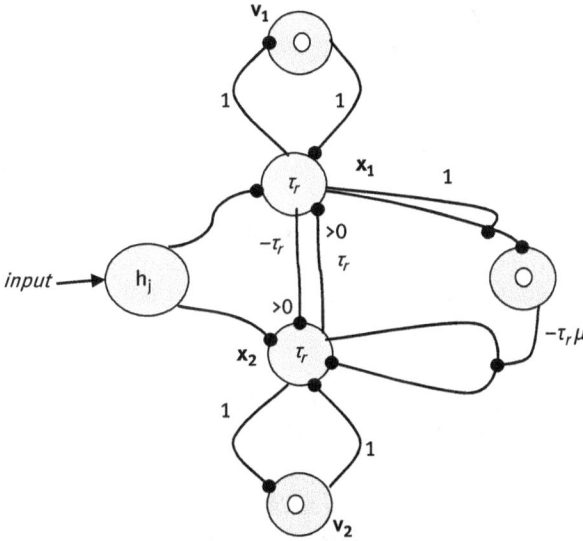

Fig. 2. VDPO model using notation in [10].

For instance, Fig. 2 shows the RNN circuit for the Van der Pol equation, which is be made by using 2 neurons x_1 and x_2. The input is taken through a neuron with a time delay h_i for smoothing reason.

Let $p = t_{r1} - t_{ro} - \dfrac{1}{\omega_r}$ be the time duration when the robot stays at the maximum rolling, where t_{r1} is the first switching time at the single support phase, then we can write $f_1 = \dfrac{1}{T}\left(\dfrac{1}{\omega_r} + p\right)$ and $f_2 = 2f_1 + \dfrac{1}{\omega_r T}$.

2.3 Van der Pol Oscillator Generating Rolling Motion

Knowing the phase portrait of the ZMP in the lateral plan can inspire the type of the controller that may substitute the piecewise linear pattern generator and therefore reduce all manual setting of the parameters as well as enhance a compact controller that can be implemented on a small microcontroller. One of the oscillators that can exhibit stable limit cycle is the Van der Pol Oscillator (VDPO), which can be described by a second order nonlinear differential equation (8) and could be regarded as a mass-spring-damper system.

$$\ddot{x} + \mu(x^2 - 1)\dot{x} + x = f(t) \tag{8}$$

where x and $\dot{x}$ represents the states of the system and μ is the damping parameter, which represents the degree of nonlinearity of the system. Function $f(t)$ represents the input from the system dynamics (effect of coupling the oscillator to the robot dynamics). Equation (8) can be re-written as

$$\ddot{x} + 2\zeta(x)\dot{x} + x = f(t) \tag{9}$$

and

$$\zeta(x) = \frac{1}{2}\mu(x^2 - 1) \tag{10}$$

$\zeta(x)$ is the damping ratio of the system and μ is the convergence rate of the system to the stable limit cycle. However, the damping factor ζ now is no more constant as it is in (1). It is a function of the state variable x of the system and therefore, stable or unstable limit cycle can be obtained depending of the value of both μ and x. Therefore, controlling the variable x and modulating the parameter μ, the oscillator limit cycle becomes adaptive to the dynamics of the robot. When there is a request to stop the locomotion of robot, the damping ζ is negative value by fixing the value of x at a specific value x and $\zeta = \frac{1}{2}\mu(x^2 - 1)$.

Let us now consider that the robot is controlled by a VDPO. Also let us consider the robot dynamics when hopping between the two legs as a hardware oscillator. The total system, therefore, can be described as two coupled oscillators expressed by (11) and (12).

$$\ddot{x}_{osc} + 2\zeta(x_{osc})\dot{x}_{osc} + x_{osc} = f(x_{zmp}, \dot{x}_{zmp}, \tau) \tag{11}$$

$$\ddot{x}_{zmp} + 2\zeta_r(x_{zmp})\dot{x}_{zmp} + x_{zmp} = f(x_{osc}, \dot{x}_{osc}, \tau) \tag{12}$$

where x_{zmp} is the calculated lateral ZMP using the force sensors located under the sole plates of the robot, ζ_r is damping factor of hardware (robot), and τ is a time dealy.

Let the coupling function be a proportional derivative controller as,

$$f(x_{zmp}, \dot{x}_{zmp}, \tau) = k_p x_{zmp}(t - \tau) + k_v \dot{x}_{zmp}(t - \tau) \tag{13}$$

where k_p and k_v are the coupling gains. τ is the time delay caused by filtering the force sensors' signals used for calculating the x_{zmp}. The time delay was decided by investigating the frequency spectrum of the force sensor output. In the experiment, the value of the time delay was set as 0.01ms. It is important to notice that gains k_p and k_v are tuned to have smooth effect on the posture of the robot that take effect along 4 to 5 gait duration. Indeed the robot's initial posture is set such that the ZMP position is almost in the middle of the supporting polygon. During locomotion of the robot, the ZMP is subject to a backward or forward deviation depending to the movement of the humanoid robot. This affects the robot posture to bend to the front or to back, respectively. This bending to the fall of the robot usually takes place during 4 to 5 times the walking cycle duration (4–5 seconds). Therefore, with the ZMP controller this bending is compensated by correcting the posture for a given desired ZMP position.

On the other hand, the damping function in (10)can be modified by introducing a parameter "*r*" that modulates the amplitude of the rolling motion regardless of the initial rolling value.

$$\zeta(x) = \frac{1}{2}\mu(x^2 - r) \tag{14}$$

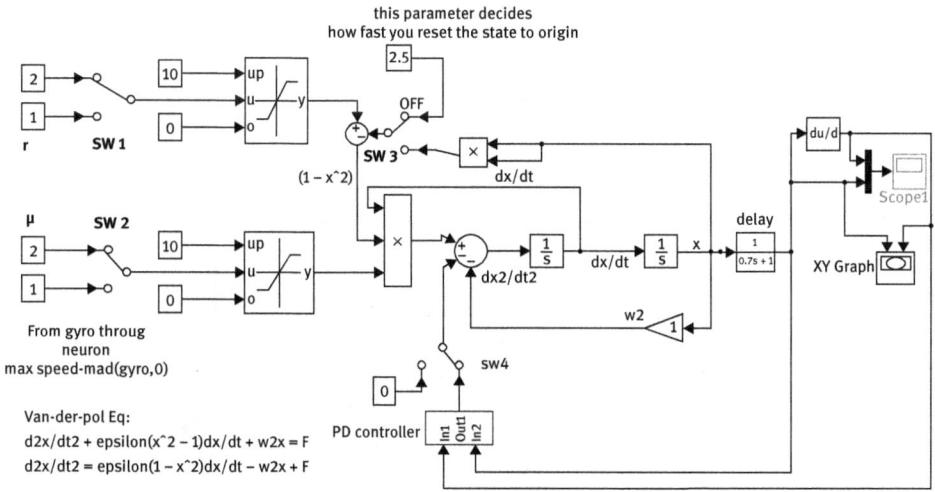

Fig. 3. VDPO model with rolling and convergence rate parameters defined in (14).

Figure 3 shows the simulation model reflecting (11) and (14) using Simulink. For instance, the rolling parameter "*r*" can be switched from 1 to 2 units. The damping parameter μ, which can speed up or slow down the convergence to the stable limit cycle, can be considered as an input from the gyro sensor through an evaluation function.

Figure 4 shows the output of the VDPO, which represents the lateral ZMP position against its change rate. It should be noticed here that with a simple derivation it can be shown that the ZMP position is directly proportional to the rolling angle of both the hip and ankle joint for small angles. It is worth noticing that Fig. 4 shows a smooth switching from a limit cycle to another limit cycle by changing the rolling parameter *r*. Such a change may take place when the robot slows down its walking speed. In such a scenario more rolling amplitude is required to maintain stability of the robot.

The convergence rate of the oscillator in Fig. 5 is $\mu = 2$, which makes the robot reach its stable walking rhythm faster than in the case when $\mu = 0.2$ (Fig. 4).

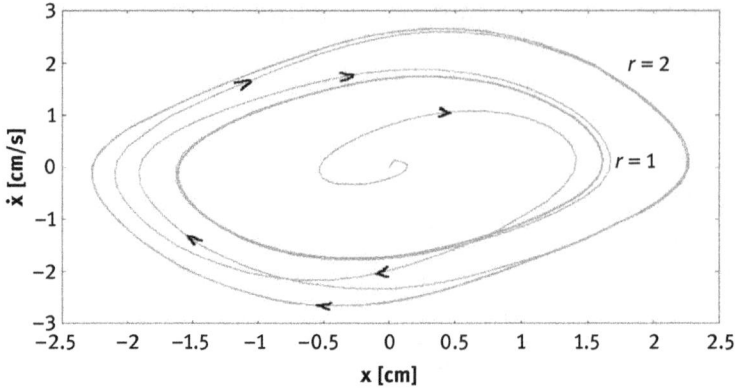

Fig. 4. Phase portrait showing the limit cycles when the rolling parameter "r" is changing from 2 to 1 with $\mu = 0.2$.

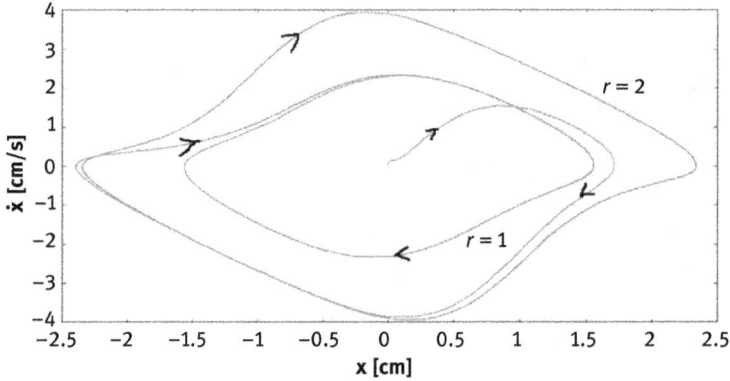

Fig. 5. Phase portrait showing the limit cycles when the rolling parameter "r" is changing from 2 to 1 with $\mu = 2$.

Sometimes the humanoid robot may need to stop suddenly at the single support phase (standing on one leg), in such a case the rolling has to be closer to its static rolling value α_s. Such case is considered in the Section 4, when the robot is stopped on one leg after being pushed from the back. Notice that the change in the rolling is done smoothly through a neuron as $\varepsilon_r \dfrac{dr'(t)}{dt} + \Delta r'(t) = \Delta r$, and the new rolling amplitude is then $r = r + r'$. Figure 6 shows the phase portrait of the state space when the stable limit cycle is switched to origin by setting "SW3" to "off" position, where $m = 2.5$, $\mu = 0.8$, $r = 2$. In contrast, Fig. 7 shows similar case but with $m = 4$, $\mu = 0.8$, $r = 2$. It is clear that when $m = 4$, the state trajectory converges faster to origin.

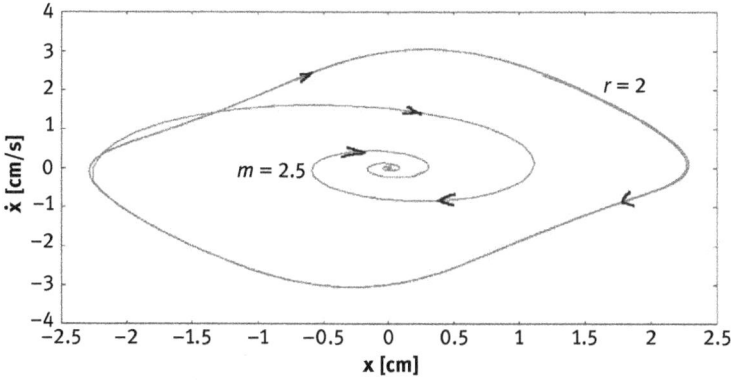

Fig. 6. Phase portrait when switching the state trajectory to origin for $m = 2.5$, $\mu = 0.8$, $r = 2$.

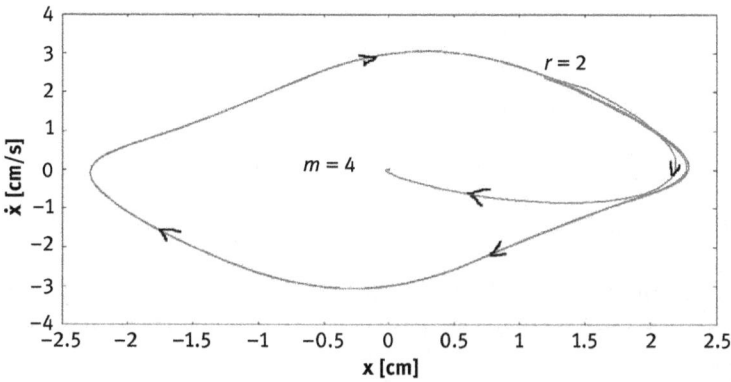

Fig. 7. Phase portrait when switching the state trajectory to origin for $m = 4$, $\mu = 0.8$, $r = 2$.

Now, if we express (11) of the VDPO in the state space form as follows:

$$\begin{cases} \dot{x}_1 & = & x_2 \\ \dot{x}_2 & = & -x_1 - 2\zeta(x_1)x_2 + f_b \end{cases} \tag{15}$$

where $x_1 = x_{osc}$ and $f_b = f(x_{zmp}, \dot{x}_{zmp}, \tau)$.

Now, let's consider how to stop the robot from walking i.e., switching the state space trajectory from the stable limit cycle to origin, one way is to set the value of "x^2" in (14) to a certain positive value "m" such that $m > r$. In this case the convergence rate to origin will depend on the total damping factor μ_T.

$$\mu_T = \mu(m - r) - k_D \tag{16}$$

Now, let the energy be $E = \frac{1}{2}(x_1^2 + x_2^2)$ then it can be shown that the change of energy during rolling for $r = 1$ can be given by;

$$\dot{E} = -2\zeta(x_1)x_2^2 + f_b x_2 \tag{17}$$

which proves that in the absence of feedback ($f_b = 0$) the system gains energy when $-r < x_1 < r$ and dissipate energy otherwise. The net exchange of energy during one oscillation should be zero to have stable closed orbit i.e., stable limit cycle. The feedback controller will play a major role in compensating for energy loss caused by friction.

2.4 Entrainment Algorithm of the Rolling Oscillator

In this section we propose an entrainment algorithm of the coupled oscillator described by (11) and (12) by resetting the PD controller at the double support phase and tuning the oscillator phase to much the dynamics phase using the measured ZMP position and velocity along the lateral plane. It should be noticed that the proposed algorithm is applicable for both PWLO and VDPO, which can be described by the following steps.

2.4.1 Entrainment Algorithm

This algorithm starts running when the humanoid robot is in its normal walking state. In other words, the state of the robot is always away from the origin and running along the stable limit cycle.

- **Step 1:** *Calculate the phase difference φ between the oscillator phase and the dynamics phase at the maximum rolling amplitude and repetitively at each walking cycle and when x_{osc} is exactly at its maximum value ($x_{osc} = r$).*

  $\varphi = \theta_{osc} - \theta_{zmp}$

  where $\theta_{osc} = \arctan \dfrac{\dot{x}_{osc}}{x_{osc}}$, $\theta_{zmp} = \arctan \dfrac{\dot{x}_{zmp}}{x_{zmp}}$, and both x_{zmp} and $\dot{x}_{zmp}$ are measured variables.

  Note that x_{zmp} never gets zero value during locomotion as it is calculated when x_{osc} is at its maximum value.

- **Step 2:** *Reset the PD controller states when $x_{osc} = r$, i.e., $x_1 = x_2 = \dot{x}_1 = \dot{x}_2 = 0$ through a neuron as in (3).*

- **Step 3:** *At $x_{osc} = r$, calculate the φ and change the phase oscillator accordingly by changing the oscillator's parameters; if φ < 0 then reduce θ_{osc}, otherwise increase θ_{osc}. The rate of change is set experimentally and varying between 0.1 and 0.2 at each walking cycle.*

- **Step 4:** *Stop tuning if the phase difference φ is within the desired range, otherwise restart over from step 1.*

3 Pitching and Swing Pattern Generator with Feedback Controller

The dynamic of the swing leg can be considered the same as that of a pendulum, and hence it is inherently stable without any compensator. Due to the need of switching control, this motion is to be generated using PWL function as shown in Figs. 8 and 9.

$$\varepsilon \frac{d\theta_l(t)}{dt} + \theta_l(t) = A_l[u(t_{lo}, \omega_l) - u(t_{l1} + nT, \omega_l) + u(t_{l2} + nT, \omega_l) - u(t_{lf}, \omega_l)] \quad (18)$$

where θ_l is the lifting motion, A_l is the amplitude of lifting, and t_{lo} and t_{l2} are the switching time for the first and second lifting, t_{l1} and t_{lf} are the switching times for the first and the last landing, and ω_l is the joints' angular velocity generating the lifting motion.

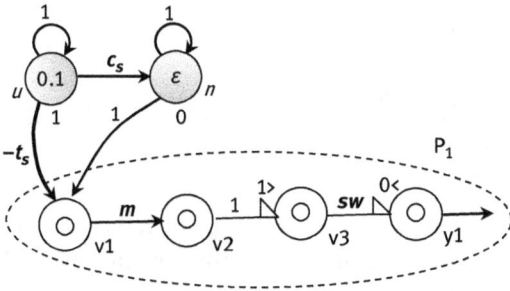

Fig. 8. Neural Circuit for pitching motion using notation in [10].

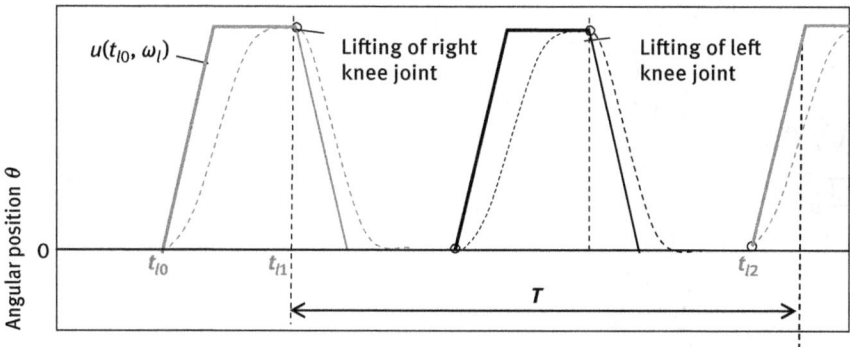

Fig. 9. Output generated by the circuit in Fig. 8. Dashed line is the smoothed output [6].

On the other hand, we assume that the landing of each leg is accomplished with flat foot on a flat ground, and the thigh and shank of the robot have the same length. Therefore, with respect to the angles definition in [11], this condition can be satisfied as follows.

$$
\begin{cases}
\theta^p_{am} &= \theta_l(t) + \theta^l_{fb}(t) + \theta_s(t) + \theta^y_{zmp}(t) \\
\theta_{km} &= -2\theta_l(t) \\
\theta^p_{hm} &= \theta_l(t) + \theta^l_{fb}(t) - \theta_s(t) - \theta^y_{zmp}(t)
\end{cases}
\tag{19}
$$

where θ^p_{am}, θ_{km} and θ^p_{hm} are the pitching motor commands to the ankle, the knee and the hip, respectively. θ^l_{fb} is the gyro feedback input satisfying the stability of inverted pendulum in the sagittal plane as shown in Fig. 10. θ^y_{zmp} is the feedback output that controls the robot posture during walking. The input of the ZMP controller is the ZMP position that is calculated using the outputs of the force sensors located under the legs (Fig. 11) as follows:

$$
x_m = \frac{x_1(F^r_1 + F^r_3 - F^l_2 - F^l_4) + x_2(F^r_2 + F^r_4 - F^l_1 - F^l_3)}{\sum\limits_{i=1}^{4}(F^r_i + F^l_i)}
\tag{20}
$$

$$
y_m = \frac{x_1(F^r_3 + F^r_4 + F^l_3 + F^l_4) + x_2(F^r_1 + F^r_2 + F^l_1 + F^l_2)}{\sum\limits_{i=1}^{4}(F^r_i + F^l_i)}
\tag{21}
$$

where x_m and y_m are the coordinates of the position of the ZMP at which the reaction torques are null. To keep the upper body non-oscillating while stabilizing the inverted pendulum $\theta^l_{fb}(t)$ is fed to the ankle and the hip as in (19).

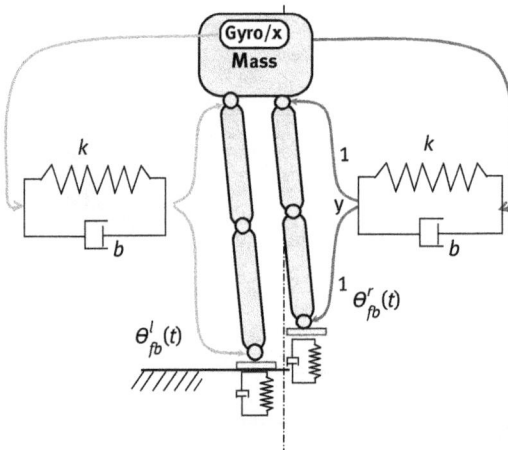

Fig. 10. Single support phase of humanoid.

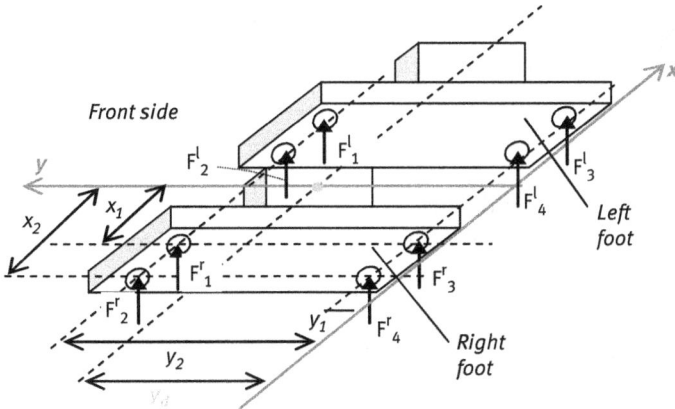

Fig. 11. Sole reaction forces on the feet and ZMP.

For the sake of unification of the design and using a similar switching law as (18), the angular position θ_s that generates the stride, can be expressed by the following equation:

$$\varepsilon\frac{d\theta_s(t)}{dt} + \theta_s(t) = A_s[u(t_{s1} + nT, \omega_s) - u(t_{s2} + nT, \omega_s)] \tag{22}$$

where A_s is the amplitude of the angle generating the stride. The t_{s1} and t_{s2} are the switching times at the start and the end of stride. The ω_s is the joints' angular velocity.

4 Reflex Against Large Disturbance

The aim of implementing reflex when a large disturbance occurs consists of preventing the humanoid robot from falling down by generating a particular motion. In this case the feedback controller will be no more sufficient to prevent the falling. The idea of doing this is to reset the rolling value r to its static rolling value, which is the needed rolling to stabilize the robot at the single support of the robot (i.e., standing on one leg). Notice that, with the presence of the PD controller, it is expected that the ZMP position will converge smoothly to the new equilibrium point.

Now, to detect a large disturbance and eventually to trigger the reflex, we use the gyro sensor. For this, we propose a normalized angular velocity index g_i in the (x, y) plane as follows:

$$g_i = \sqrt{\rho^2 gyro_x^2 + gyro_y^2} \tag{23}$$

where $gyro_x$ and $gyro_y$ are the rolling and pitching components of the gyro sensor's output along x-axis and y-axis, respectively. The parameter $\rho = \dfrac{L}{w}$, which is the size ratio of the sole plate, with L and w are the length and the width of the robot sole plate, respectively. The decision about the presence of large disturbance is based on a threshold that is determined experimentally. The ZMP location of HOAP-3 during stepping motion and in absence of large disturbance will be calculated using (20) and (21). The disturbance is considered large if the gyro index exceeds a given threshold T_{gi}, which corresponds to the critical state of the robot beyond which the stability by PD controllers is not possible.

5 Experiment

For the experiment, the humanoid robot HOAP-3 [11] of Fujitsu has been used, which has 28 degrees of freedom and is 60 cm tall and weights 8.8kg. The real-time control algorithms are implemented in real-time threads running in the RT-Linux kernel space as shown in Fig. 12. Kernel mode shared memory (SM) is constructed for the communication between real-time threads. The control period is 1 ms, and the interface between the motion pattern generator and the robot uses a real-time USB driver thread. Using the proposed locomotion controller in [6], the output of the joints angles at the hip, knee, and ankle joints are shown in Figs. 13–15 respectively, which could be produced using the VDPO as described in Figs. 4–7. The dashed line represents the output of the locomotion controller whereas the solid line represents the actual output which is adjusted by the action of the compliance controller that gets input from the sole plate sensor as shown in Fig. 10, where the damper spring model is used to emulate a flexible leg.

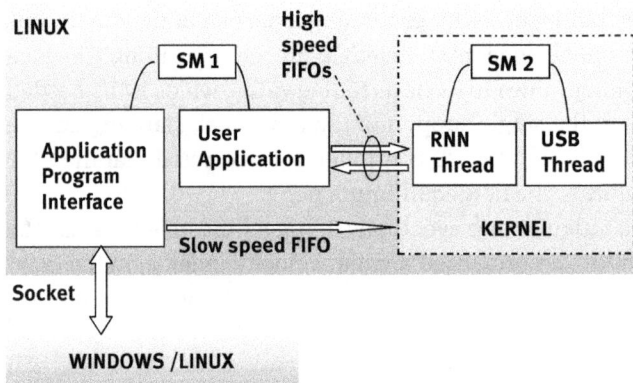

Fig. 12. Single support phase of humanoid.

Figure 16 shows the phase portrait of the ZMP position along the lateral plane and using PWLO. Regarding the case of using the VDPO, unfortunately the experiment result is not currently available and similar result is expected. A new robot is under construction where the oscillator will be implemented first on a hopping robot(i.e. monopod robot).

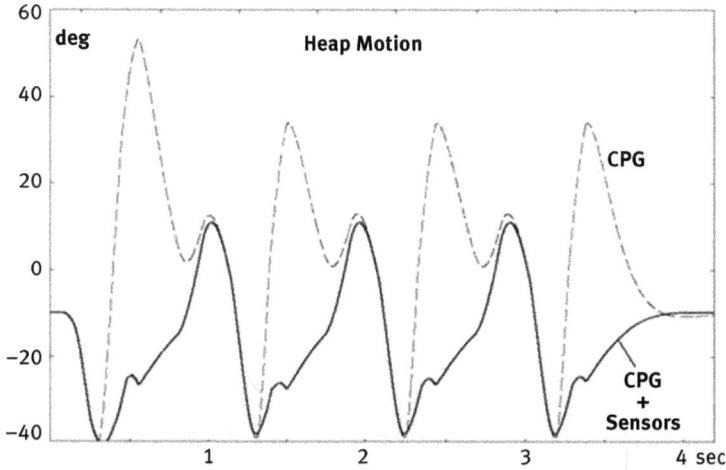

Fig. 13. Pitching motion profile of the hip joint.

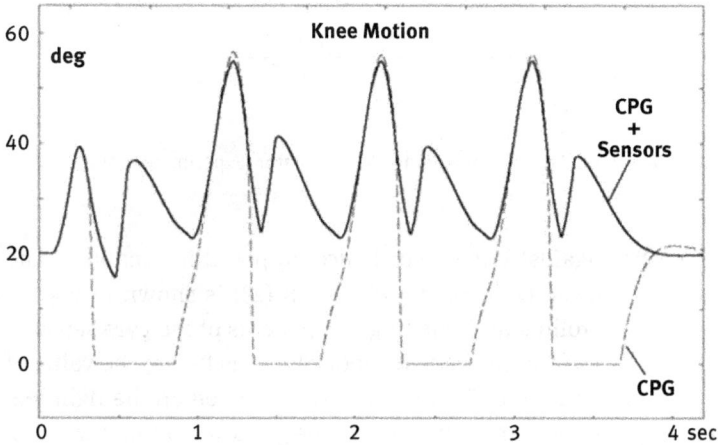

Fig. 14. Pitching motion profile of knee joint.

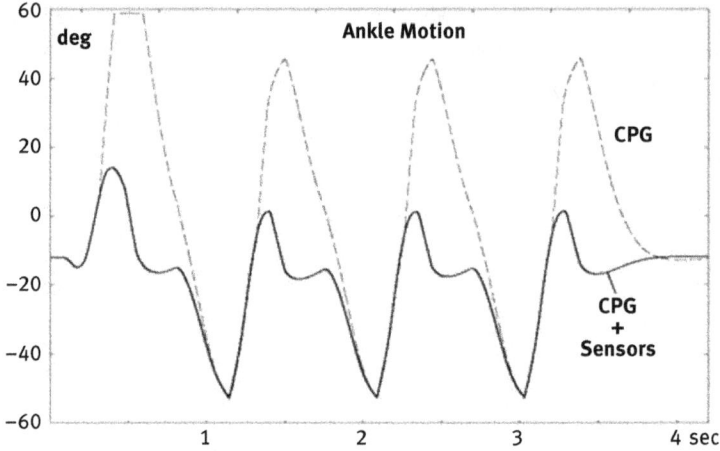

Fig. 15. Pitching motion profile of ankle joint.

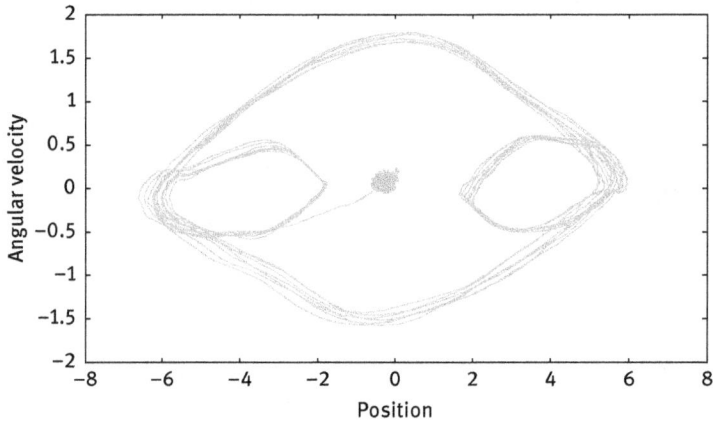

Fig. 16. Phase portrait of the ZMP in the lateral plane using PWL oscillator (Experiment result).

Let's consider now the reflex against large disturbance proposed in section 4. The normalized angular velocity in the (x, y) plane defined in (23) is shown in Fig. 17, where $gyro_x$ and $gyro_y$ are the rolling and pitching components of the gyro sensor's output. When pushing the robot from the back as noticed in this figure, the value of the gyro index gi touched the threshold T_{gi} and HOAP-3 is stopped on the right leg. $T_{gi} = 1.15 g_{imax}$, where g_{imax} is the maximum value of the g_i during normal walking as shown in Fig. 17. Notice that T_{gi} is decided experimentally. The trajectory of the ZMP x-component on the phase space is depicted in Fig. 18, which shows out how the robot was switched from a dynamical state to a static one (standing on one leg). This result, therefore, demonstrates that the switching from dynamical state (walking) to a static one was stable and smooth.

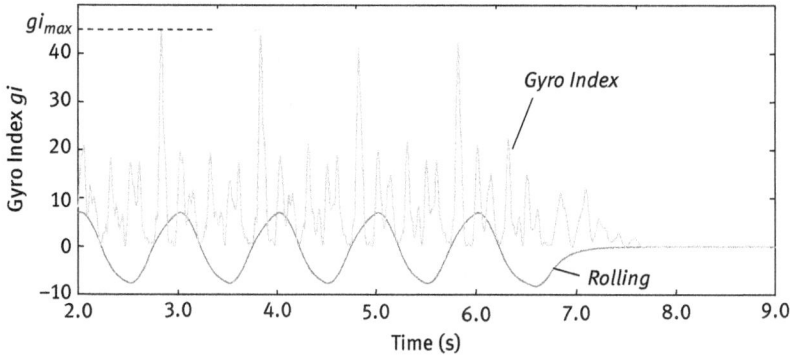

Fig. 17. Gyro index and rolling joint's output during walking when no large disturbance is present (Experiment result).

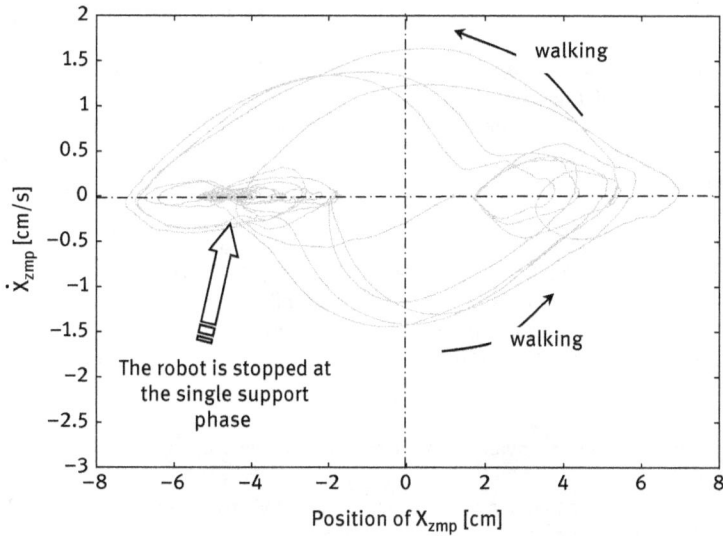

Fig. 18. Trajectory of the ZMP x-component in the phase space (Experiment result).

6 Conclusion

This paper presented a locomotion pattern generator consisting of Van de Pol Oscillator as a master oscillator that controls the rolling motion. The swing motion as well as the pitching or stride could be generated by the Van der Pol oscillator that can be considered as slave oscillators. As a result, a pretty compact solution generating rolling motion profile was obtained. By introducing a small perturbation parameter into the ODE of the master oscillator's equation, the convergence rate to the stable limit cycle could be modulated. In addition, a second parameter has been introduced representing the rolling amplitude, which could be controlled through a ZMP position. It was shown by simulation using the Van der Pol oscillator that the stable limit cycle was obtained,which is very similar to the one obtained by experiment using the humanoid robot HOAP-3. By modulating the oscillator's equation using the input from the ZMP it is possible to deal with reflex against sudden change in load and large disturbance acting on the humanoid robot upper body.

Acknowledgment: The experimental work has been conducted at Fujitsu Laboratories Ltd in Japan.

Bibliography

[1]   N. Jiangsheng, H. Seiji and K. Atsuo. A Model of Neuro-Musculo-Skeletal System for Human Locomotion Under Position Constraint Condition. *J. of Biomechanical Engineering*, 125(4):499–506, 2003.

[2]   M. Marilyn. Central Pattern Generation of Locomotion: A Review of the Evidence. *J. of the American Physical Therapy*, 82(1):69–83, 2002.

[3]   K. Matsuoka. Sustained oscillations generated by mutually inhibiting neurons with adaptation. *J. of Biological Cybernetics*, 52:367–376, 1985.

[4]   W. Lohmiller and J.J. Slotine. On contraction analysis for non-linear systems. *Automatica*, 34(6):683–696, 1998.

[5]   A. Kale, G. Salunke, K. Kadam and M. Jagdale. CPGs Inspired Adaptive Locomotion Control for Hexapod Robot. *Int. J. of Engineering Science Invention*, 2(4):2319–6734, 2013.

[6]   R. Zaier and S. Kanda. Piecewise-Linear Pattern Generator and Reflex System for Humanoid Robots. *IEEE Int. Conf. on Robotics and Automation*,:2188–2195, 2007.

[7]   K. Nakada, Y. Sato and K. Matsuoka. Tuning Time Scale Parameter of Piecewise Linear Oscillators for Phase Resetting Control. 21st *Annual Conf. of the Japanese Neural Network Society*, 2011.

[8]   P. Carla and M. Tenreiro. Complex order van der Pol oscillator. *Nonlinear Dynamics*, 65(3):247–254, 2010.

[9]   H. Hermann. Towards a unifying model of neural net activity in the visual cortex. *J. of Cognitive Neuro dynamics*, 1(1):15–25, 2007.

[10] R. Zaier and F. Nagashima. Recurrent neural network language for robot learning. 20th *Annual Conf. of the Robotics Society of Japan*, 2002.

[11] Y. Murase, Y. Yasukawa, K. Sakai and M. Ueki. Design of Compact Humanoid Robot as a Platform. 19th *Annual Conf. of the Robotics Society of Japan*,:789–790, 2001.

[12] F. Nagashima. A Motion Learning for a Robot using CPG/NP. 20th *Conf. of Robotics Society of Japan*, 2002.

Biography

Riadh Zaier received his M.Eng. and Dr. Eng. Degrees, both in Discrete-Time Tracking Control Systems from the Department of Systems Engineering, Nagoya Institute of Technology, Japan, in 1996 and 1999, respectively. He joined Fujitsu Automation Ltd. in 1999. Then he joined Fujitsu Laboratories Ltd., as a researcher in 2004. He has been conducting a leading research on the biologically inspired locomotion control for humanoid robots. He developed a neural network that generates robust and smooth motions for Fujitsu humanoid robot "HOAP", which is commercially available. He has published several patents and 4 granted US patents. He is currently a faculty member of Sultan Qaboos University and responsible for the curriculum assessment and development of the Mechatronics Engineering Program. His research interests are in the broad area of intelligent sensing and information fusion and control systems.

M. Chetoui, R. Malti, M. Aoun, M. Thomassin, M. N. Abdelkrim and
A. Oustaloup

Continuous-Time System Identification with Fractional Models from Noisy Input and Output Data using Fourth-Order Cumulants

Abstract: This paper considers the problem of identifying continuous-time fractional systems from noisy input/output measurements. Firstly, the differentiation orders are fixed and the differential equation coefficients are estimated using an estimator based on Higher-Order Statistics: fractional fourth-order cumulants based least squares (*ffocls*). Then, the commensurate order is estimated along with the differential equation coefficients. Under some assumptions on the distributional properties of additive noises and the noise-free input signals, the developed estimator gives consistent results. Hence, the noise-free input signal is assumed to be non gaussian, whereas the additive noises are assumed to be gaussian. The performances of the developed algorithm are assessed through a numerical example.

Keywords: System identification, fractional differentiation, errors-in-variables, Higher-Order Statistics, fourth-order cumulants, commensurate order.

1 Introduction

In the classical continuous-time system identification problem with fractional models, it is assumed that the system input is perfectly known while only the system output is measured with additive noise. Thus, two different classes of identification methods are developed: the first one, based on an equation error, consists in fixing the fractional differentiation orders *a priori* and estimating only the coefficients. The second one, based on an output error, makes it possible to estimate both the fractional derivatives and the coefficients. For an overview of such methods, refer to [1–9].

Systems or models where error or measurement noise affects the input and the output signals is called errors-in-variables (EIV) system. The problem of identifying such a system from noisy input and output data and consistently estimating their parameters

M. Chetoui, R. Malti, M. Aoun, M. Thomassin, M. N. Abdelkrim and A. Oustaloup: M. Chetoui[1,2], email: chetoui.manel@gmail.com, R. Malti[1], email: rachid.malti@u-bordeaux.fr, M. Aoun[2], email: mohamed.aoun@gmail.com, M. Thomassin[3,4], email: magalie.thomassin@univ-lorraine.fr, M. N. Abdelkrim[2], email: naceur.abdelkrim@enig.rnu.tn, A. Oustaloup[1], email: alain.oustaloup@ims-bordeaux.fr, [1] Université de Bordeaux, Laboratoire IMS, CNRS, UMR 5218, Talence Cedex, France, [2] Laboratoire de recherche: Modélisation, Analyse et Commande de Systèmes (MACS), Université de Gabès, Ecole Nationale d'Ingénieurs de Gabès, Tunisie, [3,4] Université de Lorraine, CRAN, UMR 7039, Campus Sciences, Vandoeuvre-lès-Nancy, France.

De Gruyter Oldenbourg, ASSD – Advances in Systems, Signals and Devices, Volume 5, 2018, pp. 125–144.
DOI 10.1515/9783110470468-008

has been investigated in the past few years. Thus, several methods have been proposed to solve the EIV system identification problem for discrete-time and continuous-time rational models. For an overview of the developed methods, refer to [10–17].

In our work, the focus is on Higher-Order-Statistics (HOS) based methods for identifying fractional linear dynamic systems in the EIV framework.

The continuous-time system identification with fractional models using third-order cumulants has been recently proposed [18–20]. The developed methods can be applied in different noisy situations: white, colored and/or mutually correlated; and give consistent estimates because of the insensitivity of the third-order cumulants to symmetrically distributed noises corrupting both input and output signals. Furthermore, the third-order cumulants based methods can be applied as long as the noise-free input signal is non-symmetric. If the latter is symmetrically distributed, its third-order cumulant equals zero. Hence, for such a process the use of fourth-order cumulants is required.

Our contribution is built around the *focls* (fourth-order cumulants based least squares) algorithm developed for continuous-time system identification with rational models in [14]. First, all differentiation orders are fixed and the fractional transfer function coefficients are estimated using the developed *ffocls* (fractional fourth-order cumulants based least squares) algorithm in different noisy situations. Then, the commensurate order is estimated along with the fractional differential equation coefficients.

The paper is organized as follows. Section 2 presents a brief review of fractional systems, a description of the EIV fractional system identification problem, some assumptions, a definition and some properties of fourth-order cumulants. The fourth-order cumulants based algorithm for continuous-time EIV system identification with fractional models is detailed in Section 3. In Section 4, a numerical study in two different noise situations: white gaussian noises on input/output and white/colored gaussian noises on input/output, illustrates the performances of the developed method. Finally, Section 5 concludes the paper.

2 Mathematical Background

2.1 Fractional Systems

A Single-Input-Single-Output (SISO) fractional Linear-Time-Invariant (LTI) system is governed by the fractional differential equation:

$$y(t) + \sum_{n=1}^{N} a_n D^{\alpha_n} y(t) = \sum_{m=0}^{M} b_m D^{\beta_m} u(t) \tag{1}$$

where $(a_n, b_m) \in \mathbb{R}^2$ are the linear coefficients of the differential equation and where $u(t)$ and $y(t)$ are respectively the input and the output signals, D is the time-domain differential operator (also denoted p); $D = \left(\dfrac{d}{dt}\right) = p$. The orders α_n and β_m are allowed to be non-integer positive numbers and are ordered for identifiability purposes:

$$0 < \alpha_1 < \ldots < \alpha_N \; ; \; 0 \le \beta_0 < \ldots < \beta_M$$

The Laplace transform of v-order fractional derivative of a time-domain function $f(t)$ relaxed at $t = 0$, i.e. $f(t) = 0$ for $t \le 0$, is [21]:

$$\mathcal{L}[D^v f(t)] = s^v F(s) \tag{2}$$

where $F(s) = \mathcal{L}[f(t)]$ and s is the Laplace variable.

Applying the Laplace transform to the differential equation (1) yields the fractional transfer function:

$$H(s) = \frac{Y(s)}{U(s)} = \frac{\displaystyle\sum_{m=0}^{M} b_m s^{\beta_m}}{1 + \displaystyle\sum_{n=1}^{N} a_n s^{\alpha_n}} \tag{3}$$

If $H(s)$ is commensurable[1] of order v, then it can be rewritten as:

$$H(s) = \frac{\displaystyle\sum_{i=0}^{m} \tilde{b}_i s^{iv}}{1 + \displaystyle\sum_{j=1}^{n} \tilde{a}_j s^{jv}} \tag{4}$$

where $m = \dfrac{\beta_M}{v}$ and $n = \dfrac{\alpha_N}{v}$ are integers and $\forall i' \in \{0, 1, \ldots, m\}, \forall j' \in \{1, \ldots, n\}$:

$$\begin{cases} \tilde{b}_{i'} = b_i \text{ if } \exists i \in \{0, 1, \ldots, M\} \text{ such that } i'v = \beta_i \\ \tilde{b}_{i'} = 0 \text{ otherwise} \\ \tilde{a}_{j'} = a_j \text{ if } \exists j \in \{1, \ldots, N\} \text{ such that } j'v = \alpha_j \\ \tilde{a}_{j'} = 0 \text{ otherwise.} \end{cases} \tag{5}$$

In rational transfer functions v equals 1 and usually numerator and denominator orders $(0 < \alpha_1 < \ldots < \alpha_N$ and $0 \le \beta_0 < \ldots < \beta_M)$ are fixed and only transfer function coefficients are estimated.

---

1 All differentiation orders are exactly divisible by the same number, an integer number of times.

The major difficulty of fractional models is time-domain simulation. Thus, two different methods for simulating fractional models are developed: the first one is based on a fractional model approximation by discrete-time rational model, like the Grünwald approximation. The second one is based on a fractional model approximation by continuous-time rational model, like the the Oustaloup approximation. For an overview of such methods, refer to [22, 23].

In this paper the CRONE Toolbox [24, 25] is used for time-domain simulation of fractional systems.

2.2 EIV Fractional System Identification Problem

Consider the fractional linear SISO system depicted in Fig. 1 with both input and output corrupted by measurement noise. The noise-free input and output signals are related by the following fractional differential equation:

$$y_0(t) + \sum_{n=1}^{N} a_n D^{\alpha_n} y_0(t) = \sum_{m=0}^{M} b_m D^{\beta_m} u_0(t) \tag{6}$$

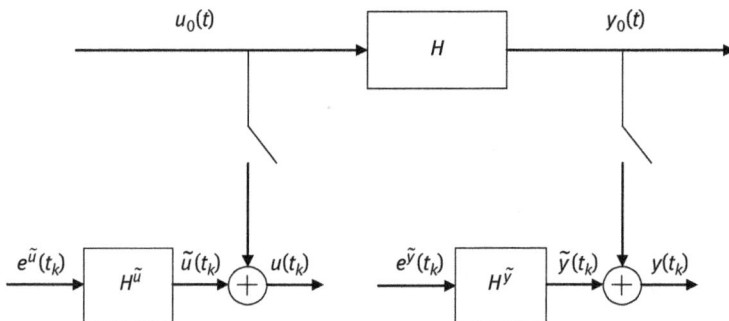

Fig. 1. Fractional system with noisy input/output data.

The continuous-time fractional system H is described by the fractional differential equation (6), $u_0(t)$ and $y_0(t)$ are the continuous-time noise-free input/output signals, $u(t_k)$ and $y(t_k)$ are the discrete-time² available measurements of input and output corrupted by additive noise $\tilde{u}$ and $\tilde{y}$.

---

2 The input/output signals are sampled uniformly at instants $\{t_k = kh\}_{k=1}^{N_t}$, where h is the sampling period and N_t is the number of measurements.

The fractional EIV linear system is thus described by:

$$\begin{cases} y_0(t) + \sum_{n=1}^{N} a_n D^{\alpha_n} y_0(t) = \sum_{m=0}^{M} b_m D^{\beta_m} u_0(t) \\ u(t_k) = u_0(t_k) + \tilde{u}(t_k) \\ y(t_k) = y_0(t_k) + \tilde{y}(t_k) \end{cases} \tag{7}$$

where

$$\begin{cases} \tilde{u}(t_k) = H^{\tilde{u}}(q)\, e^{\tilde{u}}(t_k) \\ \tilde{y}(t_k) = H^{\tilde{y}}(q)\, e^{\tilde{y}}(t_k) \end{cases} \tag{8}$$

q^{-1} denotes the backward shift operator and $e^{\tilde{u}}(t_k)$, $e^{\tilde{y}}(t_k)$ are zero-mean white gaussian noises.

The parameter vector is defined as:

$$\theta^T = [a_1, \ldots, a_N, b_0, \ldots, b_M] \tag{9}$$

The problem of identifying the EIV fractional systems is concerned with consistently estimating the parameter vector θ from N_t samples of noisy input/output data $\{u(t_k), y(t_k)\}_{k=1}^{N_t}$.

It is well known that the classical methods for system identification with fractional models give biased estimates in the EIV framework. Then, the solution is the use of Higher-Order Statistics (HOS) to get consistent estimates. Consequently, some assumptions on the distributional properties of the noisy and the noise-free signals are introduced:

- **A1:** The fractional derivative orders $0 < \alpha_1 < \alpha_2 < \ldots < \alpha_N$, $0 \le \beta_0 < \beta_1 < \ldots < \beta_M$ are a *priori* known.
- **A2:** The noise-free input signal u_0 is a zero mean stationary stochastic process such that its fourth-order cumulants are non-zero. Its probability density function (pdf) cannot therefore be gaussian.
- **A3:** $\tilde{u}$ and $\tilde{y}$ are stationary zero-mean gaussian random variables independent of u_0 and y_0. Their pdf is gaussian.

2.3 Definition and Properties of Fourth-Order Cumulants

2.3.1 Definition

The fourth-order cumulant of a real-valued, zero-mean stationary random process $\{x(t_k)\}$ is defined as [26–28]:

$$\begin{aligned} C_{xxxx}(\tau_1, \tau_2, \tau_3) &= \mathrm{Cum}\ [x(t_k)x(t_k + \tau_1)x(t_k + \tau_2)x(t_k + \tau_3)] \\ &= E\,[x(t_k)x(t_k + \tau_1)x(t_k + \tau_2)x(t_k + \tau_3)] \end{aligned}$$

$$- E[x(t_k)x(t_k + \tau_1)] \, E[x(t_k + \tau_2)x(t_k + \tau_3)]$$
$$- E[x(t_k)x(t_k + \tau_2)] \, E[x(t_k + \tau_1)x(t_k + \tau_3)]$$
$$- E[x(t_k)x(t_k + \tau_3)] \, E[x(t_k + \tau_1)x(t_k + \tau_2)] \qquad (10)$$

where Cum denotes the cumulant of $\{x(t_k)\}$ and $E[.]$ stands for mathematical expectation.

2.3.2 Properties

Higher-Order Statistics (HOS) have many properties [26–28]. Only those used in the proposed algorithm are recalled in this section. Let $x = [x(t_1), \ldots, x(t_n)]^T$ and $y = [y(t_1), \ldots, y(t_n)]^T$ be two random vectors.

-   **P1. Multilinearity:** the cumulants are linear with respect to each of their arguments. If γ_i, $i = 1, \ldots, m$, are scalars and x_i, $i = 1, \ldots, m$, are random variables, then:

$$\text{Cum } [\gamma_1 x_1, \ldots, \gamma_m x_m] = \left(\prod_{i=1}^{m} \gamma_i \right) \text{Cum } [x_1, \ldots, x_m] \qquad (11)$$

For fourth-order cumulants, $m = 4$.

-   **P2. Additivity:** if x and y are independent, the cumulant of their sum equals the sum of their cumulant:

$$\text{Cum } [x(t_1) + y(t_1), \ldots, x(t_n) + y(t_n)] =$$
$$\text{Cum } [x(t_1), \ldots, x(t_n)] + \text{Cum } [y(t_1), \ldots, y(t_n)] \qquad (12)$$

-   **P3.** The fourth-order cumulant of a random variable with a gaussian pdf equals zero.

Using properties P2 and P3, the fourth-order cumulant of the measured input/output signals is given as:

$$C_{uuuu}(\tau_1, \tau_2, \tau_3) = C_{u_0 u_0 u_0 u_0}(\tau_1, \tau_2, \tau_3)$$
$$C_{yyyy}(\tau_1, \tau_2, \tau_3) = C_{y_0 y_0 y_0 y_0}(\tau_1, \tau_2, \tau_3)$$

In a more general case, the obtained result is true for a fourth-order cumulant of any combination of input/output signals.

3 Fourth-Order Cumulants based Least Squares Algorithm

Proposition: The fourth-order cross-cumulant between the measured input and the measured output signals satisfies:

$$C_{uyuu}(\tau_1, \tau_2, \tau_3) = \frac{\sum\limits_{m=0}^{M} b_m p^{\beta_m}}{1 + \sum\limits_{n=1}^{N} a_n p^{\alpha_n}} C_{uuuu}(\tau_1, \tau_2, \tau_3) \tag{13}$$

where the differential operator p^{α_n} stands for $\dfrac{\partial^{\alpha_n}}{\partial \tau_1^{\alpha_n}}$ and p^{β_m} stands for $\dfrac{\partial^{\beta_m}}{\partial \tau_1^{\beta_m}}$, and so, verifies the following fractional differential equation:

$$C_{uyuu}(\tau_1, \tau_2, \tau_3) + \sum_{n=1}^{N} a_n p^{\alpha_n} C_{uyuu}(\tau_1, \tau_2, , \tau_3) = \sum_{m=0}^{M} b_m p^{\beta_m} C_{uuuu}(\tau_1, \tau_2, \tau_3) \tag{14}$$

Proof: See [29].

In practice, the fourth-order cumulants are estimated from N_t samples of input/output available data $\{u(t_k), y(t_k)\}_{k=1}^{N_t}$ by replacing the mathematical expectations by sample averages [27]:

$$\hat{C}_{xxxx}(\tau_1, \tau_2, \tau_3) = \frac{N_t + 2}{N_t(N_t - 1)} \sum_{k=1}^{N_t} x(t_k)x(t_k + \tau_1)x(t_k + \tau_2)x(t_k + \tau_3)$$

$$- \frac{3}{N_t(N_t - 1)} \sum_{k=1}^{N_t} x(t_k)x(t_k + \tau_1) \sum_{k=1}^{N_t} x(t_k + \tau_2)x(t_k + \tau_3)$$

$$- \frac{3}{N_t(N_t - 1)} \sum_{k=1}^{N_t} x(t_k)x(t_k + \tau_2) \sum_{k=1}^{N_t} x(t_k + \tau_1)x(t_k + \tau_3) \tag{15}$$

$$- \frac{3}{N_t(N_t - 1)} \sum_{k=1}^{N_t} x(t_k)x(t_k + \tau_3) \sum_{k=1}^{N_t} x(t_k + \tau_1)x(t_k + \tau_2)$$

Hence, with the available data, the output error may be formulated as:

$$\varepsilon(\tau_1, \tau_2, \tau_3, \theta) = \hat{C}_{uyuu}(\tau_1, \tau_2, \tau_3) - \frac{\sum\limits_{m=0}^{M} b_m p^{\beta_m}}{1 + \sum\limits_{n=1}^{N} a_n p^{\alpha_n}} \hat{C}_{uuuu}(\tau_1, \tau_2, \tau_3) \tag{16}$$

Since the estimated cumulants $\hat{C}_{uuuu}(\tau_1, \tau_2, \tau_3)$ and $\hat{C}_{uyuu}(\tau_1, \tau_2, \tau_3)$ are unbiased, the following equation holds:

$$\lim_{N_t \to \infty} \varepsilon(\tau_1, \tau_2, \tau_3, \theta) = 0 \qquad (17)$$

Then, multiplying (16) by $\left(1 + \sum_{n=1}^{N} a_n p^{\alpha_n}\right)$ and defining the equation error e_{cum} as:

$$e_{cum}(\tau_1, \tau_2, \tau_3, \theta) = \left(1 + \sum_{n=1}^{N} a_n p^{\alpha_n}\right) \varepsilon(\tau_1, \tau_2, \tau_3, \theta)$$

$$= \left(1 + \sum_{n=1}^{N} a_n p^{\alpha_n}\right) \hat{C}_{uyuu}(\tau_1, \tau_2, \tau_3) \sum_{m=0}^{M} b_m p^{\beta_m} \hat{C}_{uuuu}(\tau_1, \tau_2, \tau_3)$$

$$= \hat{C}_{uyuu}(\tau_1, \tau_2, \tau_3) - \hat{\Phi}^T(\tau_1, \tau_2, \tau_3)\theta \qquad (18)$$

where the regression vector is:

$$\hat{\Phi}^T(\tau_1, \tau_2, \tau_3) = \left[-p^{\alpha_1} \hat{C}_{uyuu}(\tau_1, \tau_2, \tau_3), \ldots, -p^{\alpha_N} \hat{C}_{uyuu}(\tau_1, \tau_2, \tau_3), \right.$$
$$\left. p^{\beta_0} \hat{C}_{uuuu}(\tau_1, \tau_2, \tau_3), \ldots, p^{\beta_M} \hat{C}_{uuuu}(\tau_1, \tau_2, \tau_3) \right] \qquad (19)$$

and the parameter vector θ is defined by (9).

The *ffocls* estimator $\hat{\theta}_{ffocls}$ is given by:

$$\hat{\theta}_{ffocls}(\tau_2, \tau_3, T) = \arg\min_{\theta} V(\tau_2, \tau_3, \theta, T) \qquad (20)$$

where the cost function V is defined as:

$$V(\tau_2, \tau_3, \theta, T) = \frac{1}{T} \sum_{\tau_1=0}^{T} \frac{1}{2} e_{cum}^2(\tau_1, \tau_2, \tau_3, \theta) \qquad (21)$$

and where T is a tuning parameter, its influence on the quality of estimation is underlined in the numerical example.

Finally, minimizing the last criterion with respect to θ leads to the *ffocls* estimator:

$$\hat{\theta}_{ffocls}(\tau_2, \tau_3, T) = \left[\frac{1}{T} \sum_{\tau_1=0}^{T-1} \hat{\Phi}(\tau_1, \tau_2, \tau_3) \hat{\Phi}^T(\tau_1, \tau_2, \tau_3) \right]^{-1}$$
$$\left[\frac{1}{T} \sum_{\tau_1=0}^{T-1} \hat{\Phi}(\tau_1, \tau_2, \tau_3) \hat{C}_{uyuu}(\tau_1, \tau_2, \tau_3) \right] \qquad (22)$$

To compute the time-domain derivatives of fourth-order cumulants estimates in the regression vector (19), the State Variable Filter approach (SVF) is proposed (see [30] for details). Then, the idea of generalizing the SVF approach to deal with fractional orders is proposed in [18] for third-order cumulants.

Similarly, the fractional derivatives of fourth-order cumulants estimates $p^\nu \hat{C}_{xxxx}(\tau_1, \tau_2, \tau_3)$ are substituted by filtered fractional derivatives of cumulants $p^\nu \hat{C}_{xxxx,f}(\tau_1, \tau_2, \tau_3)$ which can be evaluated according to:

$$p^\nu \hat{C}_{xxxx,f}(\tau_1, \tau_2, \tau_3) = \hat{C}_{xx_f^\nu xx}(\tau_1, \tau_2, \tau_3) \tag{23}$$

where x_f^ν is the signal x filtered through the SVF filter:

$$F_\nu(p) = p^\nu \left(\frac{\lambda}{p+\lambda}\right)^{\lfloor \alpha_N \rfloor + 1} \tag{24}$$

$\lfloor \cdot \rfloor$ stands for the floor function and where λ is the filter cut-off frequency.

Proof: See [29].

The implementation of the developed *ffocls* algorithm is summarized in three steps:

- **Step 1:** Define the fractional filter as in (24) and generate fractional derivatives of the filtered input/output measured signals:

$$u_f^{\beta_m}(t_k) = F_{\beta_m}(p) u(t_k); \ 0 \leqslant m \leqslant M \tag{25}$$

$$y_f^{\alpha_n}(t_k) = F_{\alpha_n}(p) y(t_k). \ 1 \leqslant n \leqslant N \tag{26}$$

- **Step 2:** Compute fractional derivatives of the fourth-order cumulants estimates using (15) and (23).
- **Step 3:** Build the filtered regression vector:

$$\hat{\Phi}_f^T(\tau_1, \tau_2, \tau_3) = \left[-p^{\alpha_1} \hat{C}_{uyuu,f}(\tau_1, \tau_2, \tau_3), \ldots, -p^{\alpha_N} \hat{C}_{uyuu,f}(\tau_1, \tau_2, \tau_3), \right.$$
$$\left. p^{\beta_0} \hat{C}_{uuuu,f}(\tau_1, \tau_2, \tau_3), \ldots, p^{\beta_M} \hat{C}_{uuuu,f}(\tau_1, \tau_2, \tau_3) \right] \tag{27}$$

and estimate the parameter vector $\hat{\theta}_{ffocls}$:

$$\hat{\theta}_{ffocls}(\tau_2, \tau_3, T) = \left[\frac{1}{T} \sum_{\tau_1=0}^{T-1} \hat{\Phi}_f(\tau_1, \tau_2, \tau_3) \hat{\Phi}_f^T(\tau_1, \tau_2, \tau_3) \right]^{-1}$$
$$\left[\frac{1}{T} \sum_{\tau_1=0}^{T-1} \hat{\Phi}_f(\tau_1, \tau_2, \tau_3) \hat{C}_{uyuu,f}(\tau_1, \tau_2, \tau_3) \right]$$

4 Numerical Example

The performances of the proposed algorithm are analyzed through the following second order fractional system:

$$H(s) = \frac{b_1 + b_2 s^v}{1 + a_1 s^v + a_2 s^{2v}} \tag{28}$$

where $b_1 = -1$, $b_2 = 1$, $a_1 = 2$, $a_2 = 1$ and $v = 0.5$.

This section is divided into two parts. In the first part, the differentiation orders are supposed to be known and only the linear coefficients are estimated using the *ffocls* estimator. In the second part, the commensurate order is supposed unknown and is estimated along with the fractional transfer function coefficients.

The noise-free input signal u_0 is chosen as a multi-sine signal (the assumption A_2 is satisfied):

$$u_0(t) = \sin(0.2t) + \sin(0.8t) + \sin(t) + \sin(1.4t) + \sin(2t) + \sin(3t) \tag{29}$$

The measurements of the input/output signals are sampled uniformly with a sampling period $h = 0.05$s (Fig. 2). The number of samples is $N_t = 5000$ and the filter cut-off frequency in (24) is set to $\lambda = 3rad/s$.

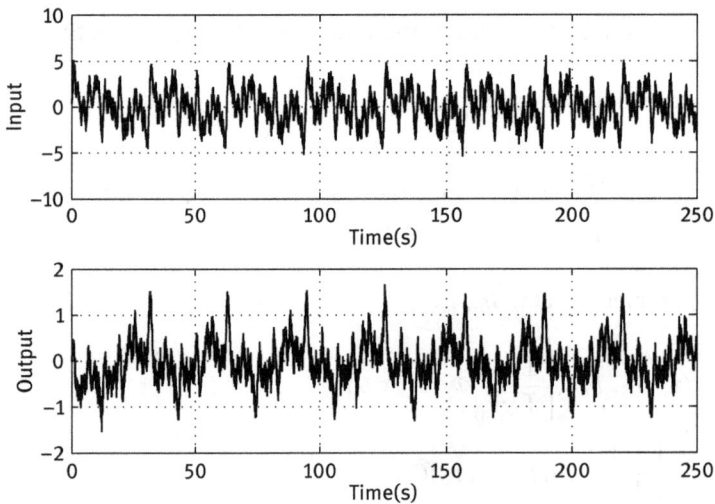

Fig. 2. Noisy input/output signals.

4.1 Known Differentiation Orders

First, the differentiation orders are considered known (the assumption A_1 is satisfied) and the fractional transfer function coefficients are estimated by applying the *ffocls* estimator in two different noisy situations.

4.1.1 White gaussian noise on input/output data

In this part the input/output signals are contaminated by additive gaussian white noise with a Signal-to-Noise-Ratio (SNR) of 10dB (the assumption A_3 is satisfied).

The *ffocls* algorithm is applied to estimate the coefficients of the fractional transfer function given by (28) for $T = 200$. The estimation results are summarized in Tab. 1.

The relative quadratic error is defined as:

$$RQE = \sqrt{\frac{\|\hat{\theta} - \theta_0\|^2}{\|\theta_0\|^2}} \qquad (30)$$

where θ_0 is the true parameter vector.

- **Influence of the tuning parameter T**: The influence of the tuning parameter T on the quality of the estimation is studied in this section. It is chosen to range between 20 and 500 and the developed *ffocls* estimator is applied for each value of T.

  The evolution of the estimates according to the tuning parameter T is plotted in Fig. 3. It is clear that for small values of T ($T < 100$) the estimates are biased, since the fourth-order cumulants are estimated using an insufficient number of samples.

- **Monte Carlo simulation**: The performances of the developed estimator are analyzed with the help of $n_{mc} = 500$ runs of Monte Carlo simulation with different gaussian white noise realizations ($SNR = 10$dB). For each Monte Carlo run, the *ffocls* algorithm is applied to estimate linear coefficients of the fractional system.

Tab. 1. Estimation results for $T = 200$: white gaussian noises on input/output.

Parameters	True Value	Estimated	RQE
a_2	1	0.9864	
a_1	2	2.058	0.0246
b_2	1	1.025	
b_1	−1	−0.9975	

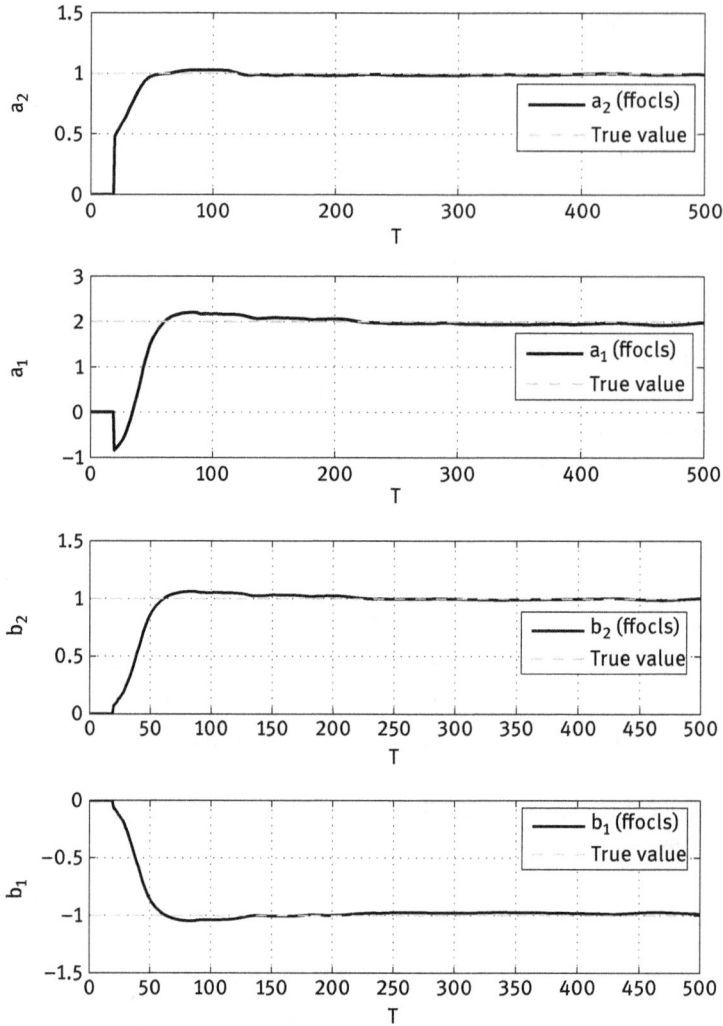

Fig. 3. Evolution of *ffocls* estimates according to the tuning parameter *T*: white gaussian noises on input/output.

The obtained results are given in Tab. 2 which contains the mean of estimates, their standard deviation (σ) and the normalized relative quadratic error.

$$NRQE = \sqrt{\frac{1}{n_{mc}} \sum_{j=1}^{n_{mc}} \frac{\left\| \hat{\theta}_j - \theta_0 \right\|^2}{\left\| \theta_0 \right\|^2}} \qquad (31)$$

It is clear that the *ffocls* estimator gives consistent estimates. Indeed, estimated coefficients converge to the real ones.

Tab. 2. Monte Carlo simulation: white gaussian noises on input/output ($T = 200$).

Parameters	True Value	Estimated	σ	NRQE
a_2	1	1.00	0.03	
a_1	2	1.94	0.14	0.064
b_2	1	0.98	0.04	
b_1	-1	-0.98	0.03	

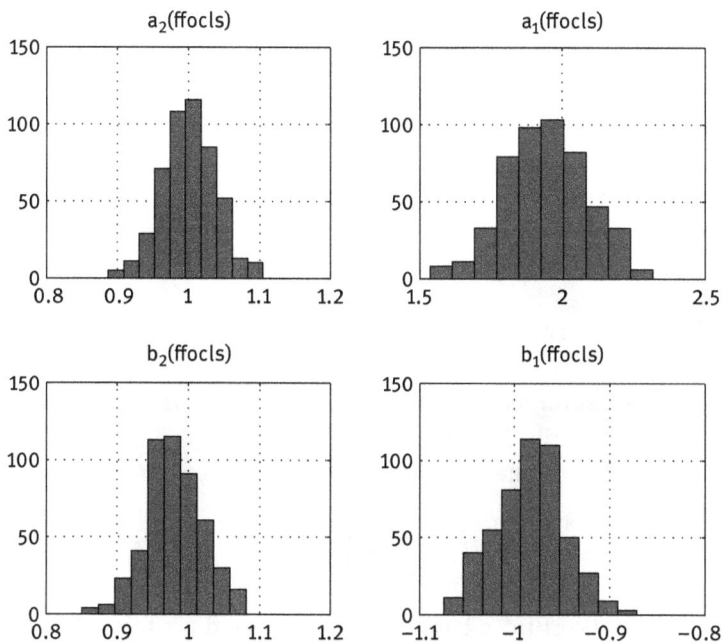

Fig. 4. Distribution of *ffocls* estimates: white gaussian noises on input/output.

The histograms of the estimates are presented in Fig. 4. It is shown that the estimates are accurate which improves the performance of the proposed algorithm.

4.1.2 White/colored gaussian noise on input/output data

In this part the noise corrupting the input and output signals are respectively white and colored and defined as:

$$\begin{cases} \tilde{u}(t_k) = e^{\tilde{u}}(t_k) \\ \tilde{y}(t_k) = \dfrac{1+0.3q^{-1}}{1-0.3q^{-1}} e^{\tilde{y}}(t_k) \end{cases} \tag{32}$$

Tab. 3. Monte Carlo simulation: white/colored gaussian noises on input/output ($T = 200$).

Parameters	True Value	Estimated	σ	NRQE
a_2	1	1.0046	0.0596	
a_1	2	1.9021	0.2223	0.1007
b_2	1	0.9676	0.0608	
b_1	−1	−0.9722	0.0555	

where $e^{\tilde{u}}(t_k)$ and $e^{\tilde{y}}(t_k)$ are realizations of white gaussian noise with a Signal-to-Noise-Ratio of 10dB.

The estimation results for $n_{mc} = 500$ runs of Monte Carlo simulation are summarized in Tab. 3. The distribution of the *ffocls* estimates are plotted in Fig. 5. It is shown that when the output noise is colored, the *ffocls* estimator still gives accurate results which improve the efficiency and the robustness to gaussian noise of the proposed algorithm.

4.2 Unknown Differentiation Orders

In this section, the commensurate order is estimated along with the coefficients in the two different noise situations described previously.

Let the l_2-norm (in dB) of the normalized output error be defined by:

$$J_{dB} = 10\log\left(\frac{\|y - \hat{y}\|^2}{\|y\|^2}\right) \tag{33}$$

where y is the measured output corrupted by an additive noise and $\hat{y}$ is the estimated output. In this section, the commensurate order is assumed to be unknown. The *ffocls* algorithms is applied to estimate the coefficients of the fractional model:

$$H(s) = \frac{b_1 + b_2 s^v}{1 + a_1 s^v + a_2 s^{2v}} \tag{34}$$

for different values of $v \in (0.1, 0.9)$ in the stability domain. The l_2-norm (in dB) is evaluated for every commensurate order v and plotted in Fig. 6. The optimum is found at $v = 0.5$ and the criteria at the optimum are close to $-$ SNR which indicates that there is no modeling error.

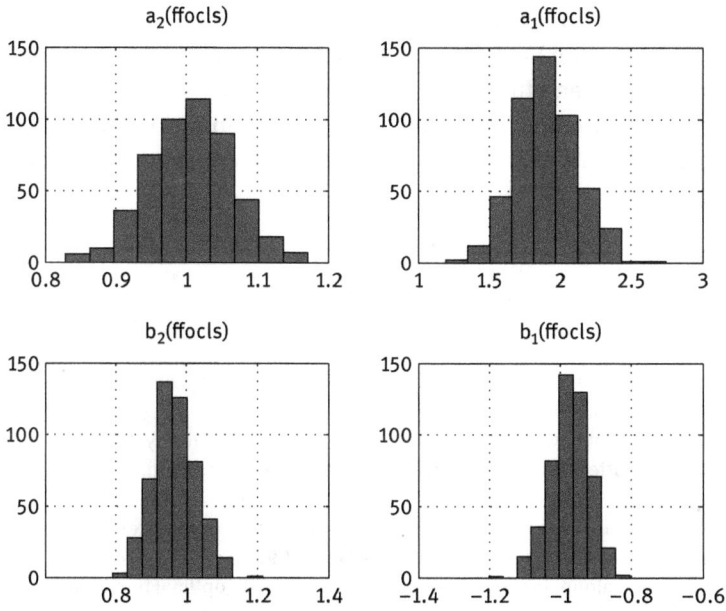

Fig. 5. Distribution of *ffocls* estimates: white/colored gaussian noises on input/output.

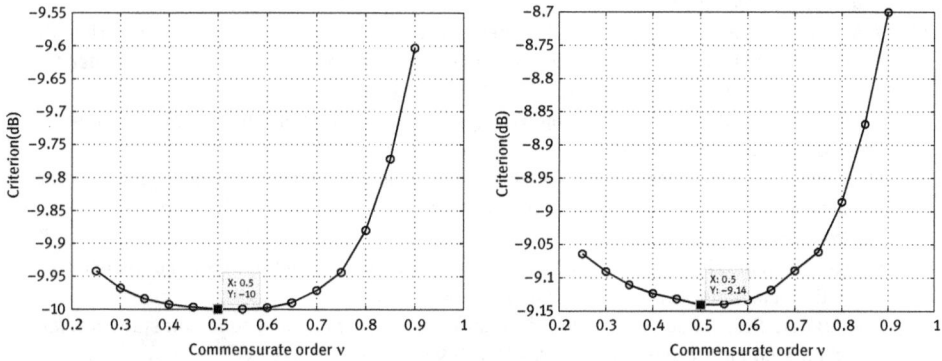

Fig. 6. Criterion versus commensurate differentiation order for white gaussian noises on input/output (left) and white/colored gaussian noises on input/output (right).

5 Conclusion

In this paper, a new continuous-time system identification method with fractional models in the EIV framework has been proposed. It is based on Higher-Order Statistics. The use of fourth-order cumulants, which are insensitive to gaussian noises corrupting both input and output signals, leads to unbiased estimates. The efficiency of the proposed estimator has been evaluated through a numerical example. The estimation results have been studied in two different noise situations: white gaussian noises on input/output and white/colored noises on input/output. Monte Carlo simulation has shown that the developed algorithm gives good results and provides unbiased estimates.

Bibliography

[1]   T. Poinot and J.-C. Trigeassou. Identification of fractional systems using an output-error technique. *Nonlinear Dynamics*, 38(1–2):133–154, 2004.

[2]   J. Sabatier, M. Aoun, A. Oustaloup, G. Grégoire, F. Ragot and P. Roy. Fractional system identification for lead acid battery state of charge estimation. *Signal Processing*, 86(10):2645–2657, 2006.

[3]   M. Thomassin and R. Malti. Subspace method for continuous-time fractional system identification. 15th *IFAC Symposium on System Identification*, 2009.

[4]   S. Victor, R. Malti and A. Oustaloup. Instrumental variable method with optimal fractional differentiation order for continuous-time system identification. 15th *IFAC Symposium on System Identification*, 2009.

[5]   M. Thomassin and R. Malti. Multivarible identification of continuous-time fractional system. *ASME 2009 Int. Design Engineering Technical Conf. & Computers and Information in Engineering Conf.*, 2009.

[6]   R. Malti, T. Raïssi, M. Thomassin and F. Khemane. Set membership parameter estimation of fractional models based on bounded frequency domain data, Communications. *Nonlinear Science and Numerical Simulation*, 15(4):927–938, 2010.

[7]   J. Gabano and T. Poinot. Fractional modelling and identification of thermal systems. *Signal Processing*, 91(3):531–541, 2011.

[8]   M. Amairi, M. Aoun, S. Najar and M. Abdelkrim. Guaranteed frequency-domain identification of fractional order systems: application to a real system. *Int. J. of Modelling, Identification and Control*, 17(1):32–42, 2012.

[9]   S. Victor, R. Malti, H. Garnier and A. Oustaloup. Parameter and differentiation order estimation in fractional models, *Automatica*, 49(4):926–935, 2013.

[10]   J. K. Tugnait. Stochastic system identification with noisy input using cumulant statistics. *IEEE Trans. on Automatic Control*, 4(37):476–485, 1992.

[11]   K. Mahata and H. Garnier. Identification of continuous-time errors-in-variables models. *Automatica*, 49(4):1470–1490, 2006.

[12]   K. Mahata. An improved bias-compensation approach for errors-in-variables model identification. *Automatica*, 43(8):1339–1354, 2007.

[13]   R. Diversi, R. Guidorzi and U. Soverini. Maximum likelihood identification of noisy input–output models. *Automatica*, 43(3):464–472, 2007.

[14]   S. Thil, H. Garnier, M. Gilson and K. Mahata. Continuous-time model identification from noisy input/output measurements using fourth-order cumulants. 46th *IEEE Conf. on Decision and Control*, :4257–4262, 2007.

[15]   S. Thil, H. Garnier and M. Gilson. Third-order cumulants based methods for continuous-time errors-in-variables model identification. *Automatica*, 44(3):647–658, 2008.

[16] S. Thil and M. Gilson. Survey of analytical iv estimates for errors-in-variables model identification. 18th *IFAC World Congress*, Milano, Italy, 2011.

[17] T. Söderström. A generalized instrumental variable estimation method for errors-in-variables identification problems. *Automatica*, 47(8):1656–1666, 2011.

[18] M. Chetoui, R. Malti, M. Thomassin, M. Aoun, S. Najar and M.N. Abdelkrim. Third order cumulants based method for continuous-time errors-in-variables system identification by fractional models. 9th *IEEE Int. Conf. on Systems, Signals and Devices*, SSD'11, Sousse, Tunisia, 2011.

[19] M. Chetoui, R. Malti, M. Thomassin, M. Aoun, S. Najar, A. Oustaloup and M. Abdelkrim. EIV methods for system identification with fractional models. 16th *IFAC Symp. on System Identification*, (SYSID), Brussels, Belgium, 2012.

[20] M. Chetoui, M. Thomassin, R. Malti, M. Aoun, S. Najar, A. Oustaloup and M. Abdelkrim. Higher-order statistics-based methods for order and parameter estimation of continuous-time errors-in-variables fractional models. 5th *Workshop on Fractional Differentiation and its Applications*, (FDA), Nanjing, China, 2012.

[21] K.-B. Oldham and J. Spanier. *The Fractional Calculus: Theory and Applications of Differentiation and Integration to Arbitrary Order*. Academic Press, 1974.

[22] M. Aoun, R. Malti, F. Levron and A. Oustaloup. Numerical simulations of fractional systems: an overview of existing methods and improvements. *Nonlinear Dynamics*, 38(1):117–131, 2004.

[23] A. Grünwald. Ueber begrenzte derivationen und deren anwendung. *Zeitschrift fur Mathematik und Physik*, 12(6):441–480, 1867.

[24] R. Malti, P. Melchior, P. Lanusse and A. Oustaloup. Object oriented CRONE toolbox for fractional differential signal processing. *Signal Image and Video Processing*, 6(3):393–400, 2012.

[25] P. Lanusse, R. Malti and P. Melchior. CRONE Control-System Design Toolbox for the control engineering community. *Philosophical Trans. of the Royal Society A: Mathematical, Physical and Engineering Sciences*, 371, 20120149, iF: 2.773, 2013.

[26] D. Brillinger. *Time series, data analysis and theory*. Holden Dat, 1981.

[27] J.-L. Lacoume, P.-O. Amblard, P. Comon. *Statistiques d'ordre Supérieur pour le Traitement du Signal*. Masson, 1997.

[28] J. Mendel. Tutorial on high-order statistics (spectra) in signal processing and system theory: Theoretical results and some applications. *Proceedings of IEEE*, 79(3):278–305, 1991.

[29] M. Chetoui, R. Malti, M. Thomassin, S. Najar, M. Aoun, M. N. Abdelkrim and A. Oustaloup. Fourth-order cumulants based method for continuous-time eiv fractional model identification. 10th *Int. Multi-Conf. on Systems, Signals and Devices*, SSD'13, 2013.

[30] P. Young. Parameter estimation for continuous-time models- a survey. *Automatica*, 17(1):23–39, 1981.

Biographies

Manel Chetoui was born in Tunisia in 1984. She received her Electrical-Automatic engineering diploma, in 2008, her Automatic and Smart Techniques Master degree, in 2009, from the ENIG (National School of Engineers of Gabes, Tunisia) and her Ph.D in 2013 in electric and automatic control from the University of Bordeaux-France and the University of Gabes-Tunisia. Her research interests include fractional differentiation, its synthesis and its applications in automatic control and system identification.

Rachid Malti was born in Serbia in 1972. He received his electrical engineering diploma from INELEC, Boumerdès, Algeria in 1994, his M.Sc and Ph.D in 1996 and 1999 in automatic control from INPL, Nancy, France. He held a position of Associate Professor at the University of Paris XII from 1999 to 2004. In 2004, he joint the IMS-CRONE team at the University of Bordeaux where he is currently holding a position of Professor in automatic control, electrical engineering, and computer engineering. Since then, he is working in the field of fractional differentiation and its applications in automatic control and system identification. He is co-developing the object oriented CRONE toolbox for fractional systems.

Mohamed Aoun was born in Tunisia in 1975. He received his Ph.D. in automatic control in 2005 from the University of Bordeaux, France. He is currently Associate Professor in automatic control, electrical engineering, and computer engineering at ENIG, Tunisia and member of its MACS research laboratory. His research interests include automatic control, system identification, fault diagnosis, and fractional differentiation.

Magalie Thomassin was born in Vitry-le-François (France) in 1978. She received the Ph.D. degree in automatic control and signal processing from the Université de Lorraine (France) in 2005. After a one-year postdoctoral fellowship at the "Laboratoire d'Automatique de Grenoble" (become the Gipsa-Lab, Grenoble, France) about the identification of open-channel flow systems, she was associate professor in the "laboratoire de l'Intégration du Matériau au Système" (IMS) at the Université de Bordeaux. She has developed new methodologies to identifying fractional order systems. Since 2010, she is associate professor at the "Centre de Recherche en Automatique de Nancy" (CRAN), Université de Lorraine, CNRS. Her research is concerned with the modeling and the identification of biological systems to control new radiation therapies.

Mohamed Naceur Abdelkrim was born in Tunisia in 1958. He obtained a Diploma in Technical Sciences in 1980, his Master Degree in Control in 1981 from the ENSET school of Tunis (Tunisia), and his PhD in Control in 1985 and the Doctorate in Sciences Degree (Electrical engineering) in 2003 from the ENIT School of Tunis. He is a Professor at the Electrical Engineering Department of the National Engineering School of Gabes (Tunisia) and he is manager of the MACS laboratory.

Alain Oustaloup was born in France in 1950. He received his engineering diploma from ENSEIRB in 1973, and his Doctor of Engineering and Doctor of Science in 1975 and 1981 from the University of Bordeaux. He is currently Professor and head of the Automatic Control department at ENSEIRB. He set up and is head of the IMS-CRONE team at the University of Bordeaux. He is currently a Research Evaluator at the French Ministry of Education and Research. His research interests include fractional differentiation, its synthesis, and its applications in engineering sciences, particularly in automatic control. He conceived CRONE control (French acronym for "Commande Robuste d'Ordre Non Entier"). He is author and co-author of five books including *fractional derivatives ant its applications in control engineering* (in French). He was awarded the Afcet Trophy in 1995 and the CNRS Silver Medal in 1997.

M. Z. E.-A. Skhiri and M. Chtourou

Generalization Improvement of Wavelet Neural Networks Using Regularization Techniques

Abstract: Similar to neural networks, the generalization of wavelet neural networks is also considered to be a serious issue since a given network may be able to assure good approximation accuracy, but could not perform well on unseen data. To improve generalization, several techniques have been carried out including regularization. In this paper, two newly regularization techniques, applied to radial wavelet neural networks, are investigated. In the first technique, the additional term of the cost function is represented in terms of a Hilbert square norm of the functional representing the network structure. However, in the second technique the network adjusted parameters decay approach is used. Applied to wavelet neural networks, this type of approach may include all the adjusted parameters and not only the weights as with neural networks.

Keywords: Radial wavelet network, generalization, regularization, square norm, adjusted network parameters decay.

1 Introduction

Learning well, it is not really learning by heart, but performing well on unseen data. That is the capability of generalization.

As a matter of fact, one of the most problems that deeply affects the performance of neural and wavelet networks is the generalization. Basically, a neural or a wavelet network may provide good approximation but could not present good performance with new data. So, the objective is to have an adequate network that presents a possible compromise between the learning performance and the generalization performance.

Similar to neural networks, the topology of wavelet neural network has also a serious impact on the performance of the network and its generalization ability. Basically, if a network does not have the adequate number of wavelets and parameters, the learning algorithm may not be able to converge. On the other hand, if the network is densely connected overfitting will take place. In other words, the network will memorize the training data instead of learning the problem, and will not be able to predict situations that it has never seen before. Statistically speaking, these two situations are related to the weight of the bias and the variance terms that compose the

M. Z. E.-A. Skhiri and M. Chtourou: Control and Energy Management Laboratory (CEMLab), National School of Engineering of Sfax, University of Sfax, Sfax, Tunisia., email: Zine.Skhiri@isetso.rnu.tn, email: Mohamed.Chtourou@enis.rnu.tn

De Gruyter Oldenbourg, ASSD – Advances in Systems, Signals and Devices, Volume 5, 2018, pp. 145–164.
DOI 10.1515/9783110470468-009

generalization error. In fact, an excessive bias will lead to under estimated network. However, an excessive variance will lead to over fitting.

To improve generalization, several techniques have been carried out, mainly, the regularization technique [1–8] which aims to prevent overfitting. This technique has been introduced in terms of several different approaches such as early stopping and weight decay.

In the early stopping method [9–15], the training set is split into two sets. One set is used for training, and the second one is used for validation. To monitor the training process, the error on the validation set is considered for the checking task. Basically, this error and the error on the training set decrease during the training process. However, when the network begins to overfit the data, the error on the validation set starts to rise. As a result, the training will be stopped and the updated parameters are returned.

In the weight decay method [16–21], however, the network complexity is reduced by limiting the growth of the weights. The basic idea is to penalize the overly large weights by including an additional term to the cost function of the learning algorithm. In order to control the strength of the regularization, a hyperparameter, known as the regularization parameter is introduced. The larger the parameter is, the higher the penalty will be on the model complexity.

In this paper, two regularization techniques are applied to radial wavelet networks. In the first technique, the additional term, called also the smoothness term, of the cost function is represented in terms of a Hilbert square norm of the functional representing the network structure. In the second technique however, the adjusted parameters decay approach is used. In fact, this method is inspired from the weight decay method.

This paper is organized as follows: In the first section, the related works are presented followed by the wavelet network structure in the second section. The adopted regularization technique and the learning algorithm are then discussed in section 4 and 5 respectively. However, the simulation results are discussed in section 6. Finally, the last section will conclude the paper.

2 Related Works

The generalization issue of neural networks has been intensively treated, and important works have been carried out using different techniques such as regularization [1–8]. The main idea of this technique is based on including an additional term to the cost function. For example, in [2] negative correlation learning NCL neural network ensemble algorithm is applied to numerous empirical problems, including regression and classification problems. NCL introduces a correlation penalty term to the cost function of each individual network so that each neural network minimizes its mean square error (MSE) together with the correlation of the ensemble. The regularization technique is also employed through other approaches such as early stopping [9–15] and weight decay [16–21]. As far as early stopping is concerned several works have also been introduced. For example, to enhance the performance of Bayesian regularization of artificial neural network a pre-training via an early-stopping algorithm is used [9]. The proposed method is applied to the regularization of Feed-forward Neural Networks to regress various benchmark data series. In a second example, early stopping is used in [10] for neural-network-based subpixel classification, which is one of the most commonly used approaches to address spectral mixture problems. A new stopping criterion has been proposed, and it is based on the reduction of the mean squared error (MSE) for a validation data set.

Due to its simplicity, weight decay technique has played also an important role to overcome overfitting, and a lot of work has been accomplished in this matter [16–21]. Based on the fact that the weight-decay method can suppress the effect of weight fault, mean prediction error (MPE) formulae are developed in [16] for predicting the performance of faulty radial basis function (RBF) networks. These MPE formulae are also able to accurately locate the appropriate value of the decay parameter for minimizing the true test error of faulty networks. Weight decay is also used in [17] for modelling nonlinear behaviour of steam temperature in distillation essential oil extraction system. The modelling is based on the neural network autoregressive with exogenous input structure. During the network training, the optimisation of the network weights has been carried out by minimisation the error through the Levenberg–Marquardt algorithm (LMA).

Similarly, for wavelet networks, different interesting works have also been carried out dealing with this issue from different points of view [1, 22–27]. Some are based on the network structure and the conventional regularization technique, and some others are treated from the signal processing point of view based on the sampling theory and the frequency band of the wavelet network. For example, in [1] the additional term of the cost function is represented in terms of a Hilbert square norm of the functional representing the network structure. In [24] however, a self-adaptive four layer wavelet network has been introduced to greatly improve generalization using the regularization technique to construct an orthogonal wavelet network based on orthogonal least squares. The generalization ability of one-dimensional wavelet networks has also been discussed in [25] from different aspects such as network complexity, over-fitting and extrapolation fitting etc... Based on the sampling theory, different algorithms have also been introduced to avoid wavelet networks overfitting. For example, in [22] the input weights are decided by the sampling period or the frequency band of target function instead of sample errors. Applying this algorithm, the wavelet network is considered to be an ideal low-pass filter, which removes the high-frequency noise in training data. For the same purpose, another algorithm robust to the variance of noise was introduced by the same author [23] based on the frequency band of the wavelet network.

In this paper two regularization techniques are applied to radial wavelet networks. It should be noticed that as far as we know the following techniques have not been treated before with wavelet networks. The idea in the first technique is based on the inclusion of an additional term to the error function. This term is expressed by a Hilbert square norm of the functional representing the network structure. In the second technique however, the adjusted parameters decay approach is used.

3 Wavelet Neural Networks

Based on the wavelet theory, Zhang and Benveniste [28] have proposed the wavelet network as an alternative to the feedforward neural networks for approximating arbitrary nonlinear functions. These types of networks are similar to the one hidden layer neural networks in which the coefficients can be considered as the connection weights and the wavelets called also wavelons as the activation functions. These wavelets are some dilated and translated versions of a mother wavelet $\psi(x)$. Compared to neural networks, wavelet networks use an important number of parameters that may be updated. They include weights, translations, dilations and direct term coefficients. Certainly, this will provide more flexibility to derive different network structures. The simple mono-dimensional input structure is represented by the following expression:

$$y(x) = \sum_{j=1}^{m} w_j \psi \left(\frac{x - t_j}{s_j} \right) \tag{1}$$

where m is the number of wavelets used in the network, and $\psi \left(\dfrac{x - t_j}{s_j} \right)$ is some translated and dilated version of the mother wavelet $\psi(x)$.

To deal with large dimensions just like the classical sigmoid neural network, an alternative network is introduced in [29, 30]. It is the radial wavelet network which is an extension of the one dimensional wavelet network. A wavelet function $\psi(x)$ is radial if it has the following form: $\psi(x) = \phi(\|x\|)$, where $\|x\| = (x^T x)^{\frac{1}{2}}$ and ϕ is a single variable function. The general expression of this network is given below in (2) including the linear direct terms.

$$y(x) = \sum_{j=1}^{m} w_j \psi [D_j(x - t_j)] + a^T x + a_0 \tag{2}$$

where D_j's are diagonal matrices built from dilation vectors: $D_j = diag(d_j)$, $d_j = \left[\dfrac{1}{s_j^1}, \dfrac{1}{s_j^2}, \dots, \dfrac{1}{s_j^N} \right]$ and $s_j = \left[s_j^1, s_j^2, \dots, s_j^N \right]^T$. N represents the input dimension. In this paper, radial multidimensional input and one dimensional output network is treated using the radial wavelet given by the following expression:

$$\psi(x) = \left(x^T x - N\right) e^{-\frac{1}{2} x^T x}, \quad x \in \mathbb{R}^N \tag{3}$$

4 Regularization Technique

Basically, the learning process of networks is performed so as to minimize the sum of squared errors between the network outputs and the training data. However, this procedure is not always able to provide to the treated network good generalization, and over fitting may occur during the training process. To improve generalization, regularization techniques are often adopted using different approaches. In this paper, two regularization techniques are applied to radial wavelet networks. For instance, we should mention that as far as we know, these two techniques have not been used before with wavelet neural networks. The first technique replaces the smoothness term by a square norm of the functional representing the network structure, and the second technique uses the network adjusted parameters decay method.

4.1 Square Norm Regularization Technique

Generally, regularization techniques are often based on the inclusion of an additional term in the error function E as stated in the following expression:

$$J[f] = E + \lambda \Omega [f] \tag{4}$$

where λ is a hyperparameter that controls the strength of the regularization and Ω is known as the regularizer. f denotes the function computed by the neural network. The first term is enforcing closeness to the data, and the second represents the smoothness. As far as neural networks are concerned, the most common used error function in training is the mean squared error (MSE):

$$E = \frac{1}{2} (y^n - f_d^n)^2 \tag{5}$$

where f_d^n and y^n are the desired output and the network output of the nth sample respectively. The smoothness term, in this case, is defined as: $\Omega [f] = \|f\|_H^2$ the square norm of f [31–33] in a reproducing kernel Hilbert space H. The kernel, in this paper, is composed of wavelets forming an orthogonal wavelet basis. The square norm in a Hilbert space is related to the inner product by: $\|f\|_H^2 = < f, f >_H$.

For instance, let $\psi_j(x)$ be an orthogonal wavelet basis, then, we may write any expansion of a function f in L_2 (integrable square functions) as follows: $f(x) = \sum_j w_j \psi_j(x)$.

Therefore, for the most simple wavelet network structure given by the following expression:

$$y(x) = \sum_j w_j \psi_j(x) = \sum_{j=1}^{m} w_j \psi\left(\frac{x - t_j}{s_j}\right) \tag{6}$$

The corresponding error may be written as follows:

$$J_1 = \frac{1}{2}(y^n - f_d^n)^2 + \lambda \|y^n\|^2 \tag{7}$$

So, the corresponding square norm will be expressed as follows:

$$\|y\|^2 = \int [y(x)]^2 \, dx = \int \left[\sum_j w_j \psi\left(\frac{x - t_j}{s_j}\right)\right]^2 dx \tag{8}$$

since

$$\left(\sum_j a_j\right)^2 = \left[\sum_j (a_j)^2\right] + 2a_1 \sum_{j>1} a_j + 2a_2 \sum_{j>2} a_j + \cdots \tag{9}$$

Equation (8) may be written as follows:

$$\int \left[\sum_j w_j \psi\left(\frac{x - t_j}{s_j}\right)\right]^2 dx = \int \sum_j w_j^2 \psi_j^2\left(\frac{x - t_j}{s_j}\right) dx +$$

$$\int 2w_1 \psi_1\left(\frac{x - t_1}{s_1}\right)\left[2w_2\psi_2\left(\frac{x - t_2}{s_2}\right) + \cdots\right] dx +$$

$$\int 2w_2 \psi_2\left(\frac{x - t_1}{s_1}\right)\left[w_3\psi_3\left(\frac{x - t_3}{s_3}\right) + \cdots\right] dx + \cdots \tag{10}$$

Provided that $\int \psi_j(x)\psi_l(x)dx = 0$ for $j \neq l$

$$\int \left[\sum_j w_j \psi\left(\frac{x - t_j}{s_j}\right)\right]^2 dx = \int \sum_j w_j^2 \psi_j^2\left(\frac{x - t_j}{s_j}\right) dx \tag{11}$$

So:

$$\|y\|^2 = \sum_j w_j^2 \int \psi_j^2\left(\frac{x - t_j}{s_j}\right) dx \tag{12}$$

Making a change of variable we get:

$$\|y\|^2 = \sum_j w_j^2 s_j \int \psi_j^2(x) dx \tag{13}$$

For orthonormal wavelets, we may write:

$$\|y\|^2 = \sum_j w_j^2 s_j \tag{14}$$

So the error function in terms of the adjusted parameters is finally given by the following expression :

$$J_1(\theta) = \frac{1}{2}(y^n - f_d^n)^2 + \lambda \sum_j w_j^2 s_j \qquad (15)$$

4.2 Adjusted Network Parameters Decay Approach

Another regularization approach is the adjusted network parameters decay that consists of summing the squares of the adjusted network parameters. Applied to neural networks, this method has the objective to penalize the overly large adjusted network parameters that cause an excessive variance term which could be controlled by the decay parameter λ_i given in equation(16). In the contrary, for small parameters the network will have an excessive bias term which could be controlled, in the same manner, by the same parameter. Therefore, the value of this parameter has a direct influence on the decay process since the generalization error is composed of bias and variance terms [34, 35]. Using this technique, the expression stated by equation (4) becomes as follows:

$$J_2(\theta) = \frac{1}{2}(y^n - f_d^n)^2 + \frac{1}{2}\sum_{j=1}^{m}(\lambda_1 w_j^2 + \lambda_2 s_j^2 + \lambda_3 t_j^2) + \frac{1}{2}\sum_{j=1}^{N}\lambda_4 a_j^2 \qquad (16)$$

5 Learning Algorithm

To adjust the parameters of the networks used in this paper, a stochastic gradient algorithm is carried out. The aim is to adjust recursively the parameters vector θ, as given in equation (17), after each input/output observation.

Vector $\theta = \left[\mathbf{w}^T \mathbf{t}^T \mathbf{s}^T \mathbf{a}^T\right]^T$ includes: the weights $\mathbf{w} = [w_1 w_2 ... w_m]^T$, the translations $\mathbf{t} = [t_1 t_2 ... t_m]^T$, the dilations $\mathbf{s} = [s_1 s_2 ... s_m]^T$ and the direct terms $\mathbf{a} = [a_1 a_2 ... a_N]^T$.

$$\theta(k+1) = \theta(k) + \Delta\theta(k) = \theta(k) - \mu\frac{dJ(\theta)}{d\theta} \qquad (17)$$

where $J(\theta)$ is given below depending on the corresponding regularization technique.
For the square norm smoothing function we have:

$$J_1(\theta) = \frac{1}{2}(y^n - f_d^n)^2 + \lambda \sum_j w_j^2 s_j \qquad (18)$$

Then:

$$\frac{dJ_1(\theta)}{d\theta} = e^n\frac{dy^n}{d\theta} + \lambda\frac{d}{d\theta}\sum_j w_j^2 s_j \qquad (19)$$

where $e^n = y^n - f_d^n$ representing the error between the network output and the desired output corresponding to the nth sample.

In the same manner, the error function using the adjusted parameters decay regularization can be given as follows:

$$J_2(\theta) = \frac{1}{2}(y^n - f_d^n)^2 + \frac{1}{2}\sum_{j=1}^{m}\left(\lambda_1 w_j^2 + \lambda_2 s_j^2 + \lambda_3 t_j^2\right) + \frac{1}{2}\sum_{j=1}^{N}\lambda_4 a_j^2 \qquad (20)$$

Then:

$$\frac{dJ_2(\theta)}{d\theta} = e^n\frac{dy^n}{d\theta} + \sum_{j=1}^{m}\lambda_1 w_j + \lambda_2 s_j + \lambda_3 t_j + \lambda_4\sum_{j=1}^{N}a_j \qquad (21)$$

At this stage, the corresponding partial derivatives will be determined. As far as the smoothness term is concerned, its partial derivative is simple and direct. However, it is more interesting to give the partial derivative of the basic cost function which is common for both regularization techniques. So, for the network expression used in this paper which is given by the following equation:

$$y(x) = \sum_{j=1}^{m} w_j\psi(D_j(x - t_j)) \qquad (22)$$

$\frac{dE(\theta)}{d\theta} = e^n\frac{dy^n}{d\theta}$ with respect to each network parameter is expressed as:

Weights w_j:

$$\frac{dE}{dw_j} = e^n\psi_j(D_j(x^n - t_j)) \qquad (23)$$

Translations t_j:

$$\frac{dE}{dt_j} = -e^n w_j D_j\psi_j'(D_j(x^n - t_j)) \qquad (24)$$

where: $\psi'(x) = \frac{d\psi(x)}{dx}$.

Dilations s_j:

$$\frac{dE}{ds_j} = -e^n w_j D_j^2 diag(x^n - t_j)\psi_j'(D_j(x^n - t_j)) \qquad (25)$$

6 Simulation Results

For instance, both regularization techniques are tested and compared to each other and to a non regularized case. Two examples will be tested using two dynamic nonlinear plants. In each case, the input will be divided into two sets. A first set is used for training, and a second one is used for checking.

6.1 Example 1

The first example is a nonlinear plant governed by the following difference equation.

$$y_d(k+1) = 1.5\frac{y_d(k)}{1+y_d^2(k)} + 0.3\cos y_d(k) + 1.2u(k) \tag{26}$$

$$\text{where: } u(k) = \begin{vmatrix} \sin\left(\dfrac{2\pi k}{250}\right), & k \le 500 \\ 0.8\sin\left(\dfrac{2\pi k}{250}\right) + 0.2\sin\left(\dfrac{2\pi k}{25}\right), & 500 < k \le 800 \end{vmatrix} \tag{27}$$

Simulations will be applied to both regularization techniques and to a third case in which no regularization term is added. For all these cases, the corresponding plant $y_d(k)$ and its estimated version denoted by $y(k)$, as well as the corresponding training and checking errors are represented in Figs. 1–6, respectively.

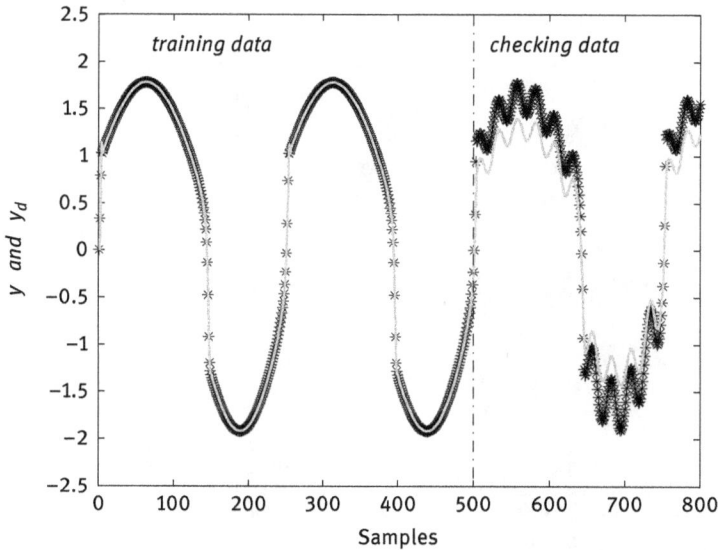

Fig. 1. The plant (dashed line) and its estimated version (continued line) with no regularization term added.

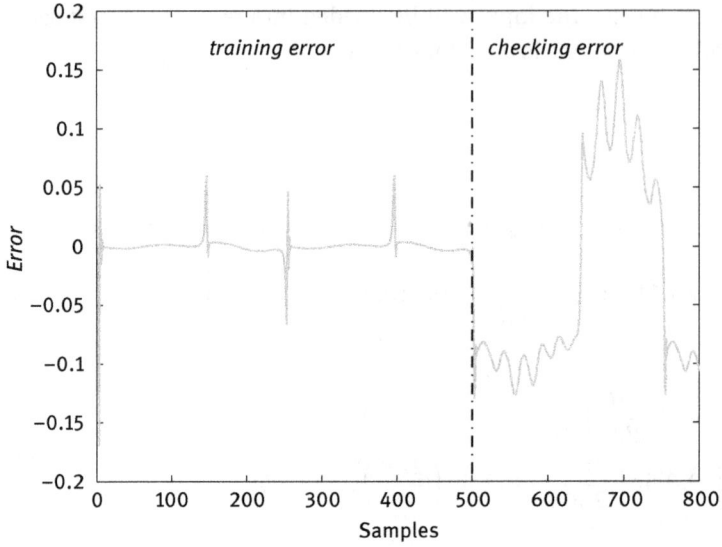

Fig. 2. Training and checking error for the case without regularization term.

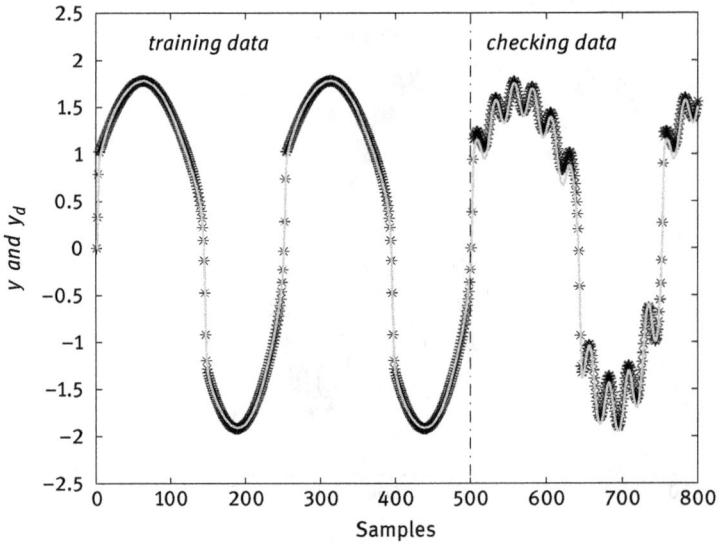

Fig. 3. The plan (dashed line) and its estimated version (continued line) using square norm regularization.

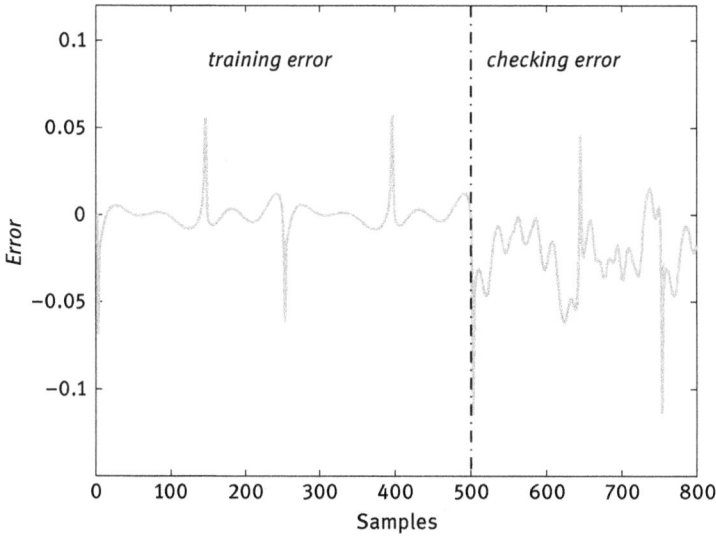

Fig. 4. Training and checking error using square norm regularization.

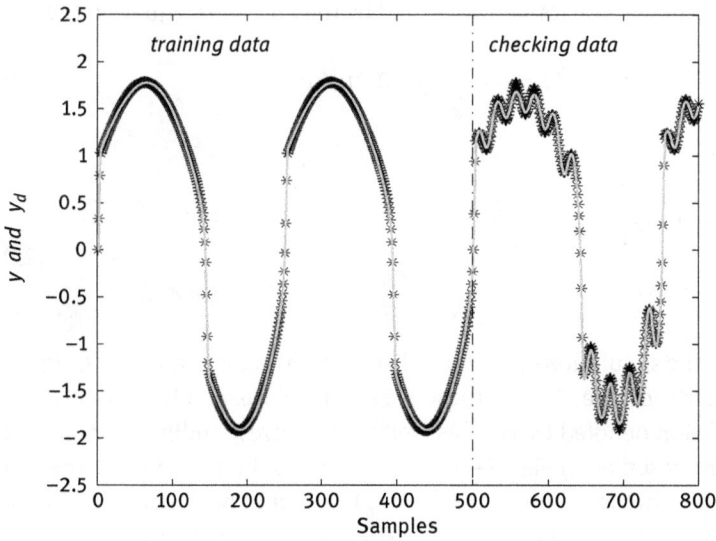

Fig. 5. The plant (dashed line) and its estimated version (continued line) using the adjusted parameters decay regularization.

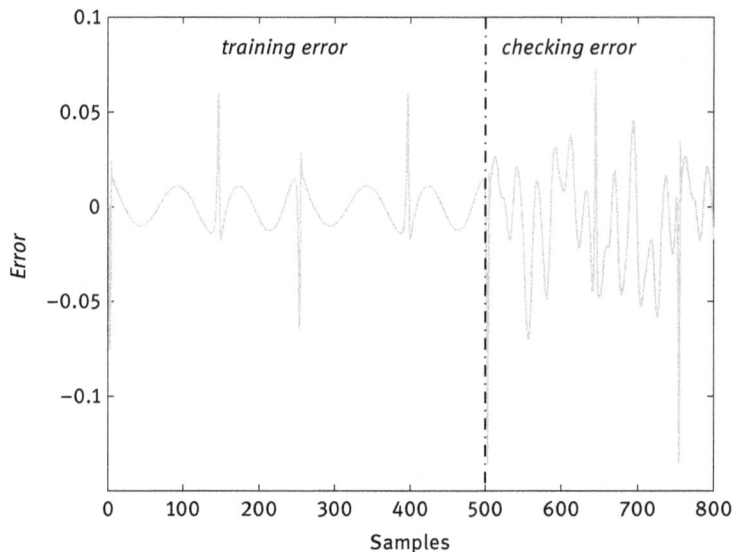

Fig. 6. Training and checking error using the adjusted parameters decay regularization.

6.2 Example 2

The second example is a nonlinear plant governed by the difference equation (28).

$$y_d(k+1) = \frac{y_d(k)}{1+y_d^2(k)} + 0.2u(k)^3 \tag{28}$$

where:

$$u(k) = \begin{vmatrix} \sin\left(\dfrac{2\pi k}{250}\right) + \sin\left(\dfrac{2\pi k}{100}\right), & k \le 500 \\[4mm] 0.8\sin\left(\dfrac{2\pi k}{250}\right) + 0.2\sin\left(\dfrac{2\pi k}{25}\right), & 500 < k \le 800 \end{vmatrix} \tag{29}$$

Similar to example 1, the simulations will be applied to both regularization techniques and to a none regularized case. For all these cases, the corresponding plant $y_d(k)$ and its estimated version denoted by $y(k)$, as well as the corresponding training and checking errors are represented in Figs. 7–12, respectively. For both examples, the rest of the training process parameters are listed in Tab. 1. It should be noticed that TMSE and CMSE stand for training and checking mean square errors respectively.

According to the simulations, it is clear that both treated approaches have guaranteed a satisfactory generalization compared to the case presenting no regularization term. However, the square norm approach has shown slightly better result compared to the adjusted parameters decay approach. This can be seen from Tab. 1, where the training and the checking mean square errors (MSE) generated by the square norm method are less than the MSE generated by the second method.

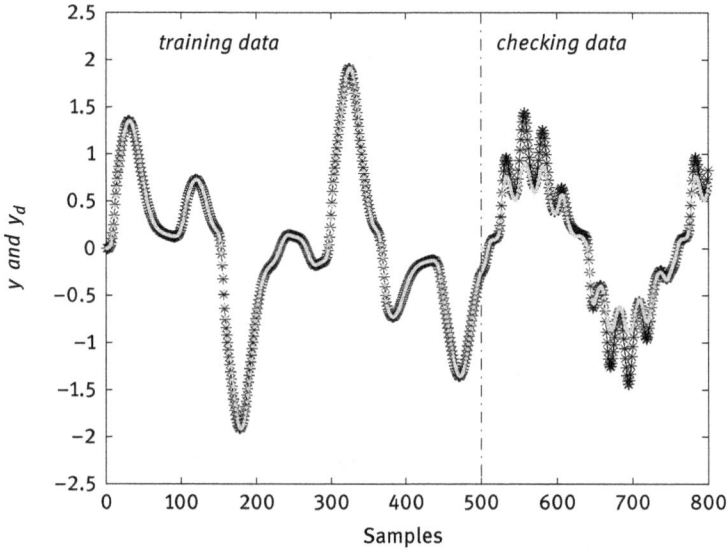

Fig. 7. The plant (dashed line) and its estimated version (continued line) with no regularization term added.

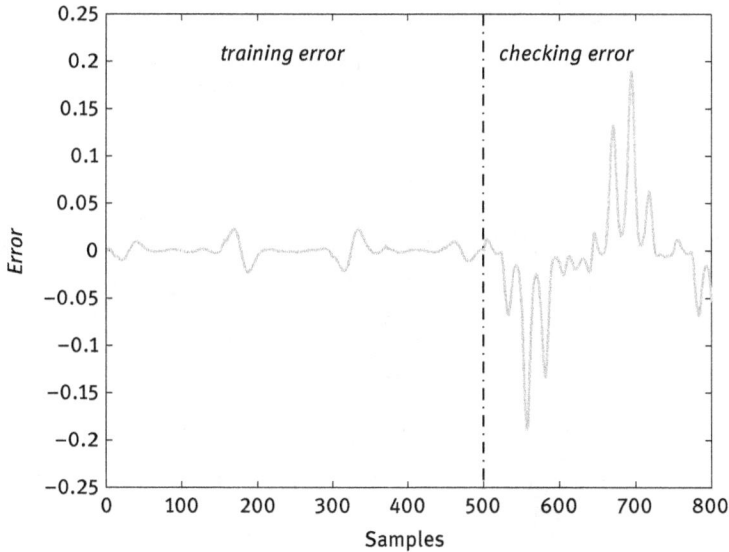

Fig. 8. Training and checking error for the case without regularization term.

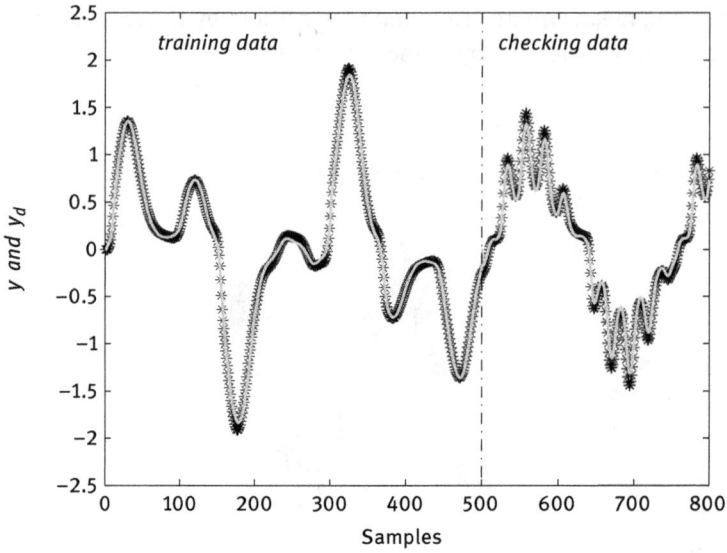

Fig. 9. The plant (dashed line) and its estimated version (continued line) using the square norm regularization.

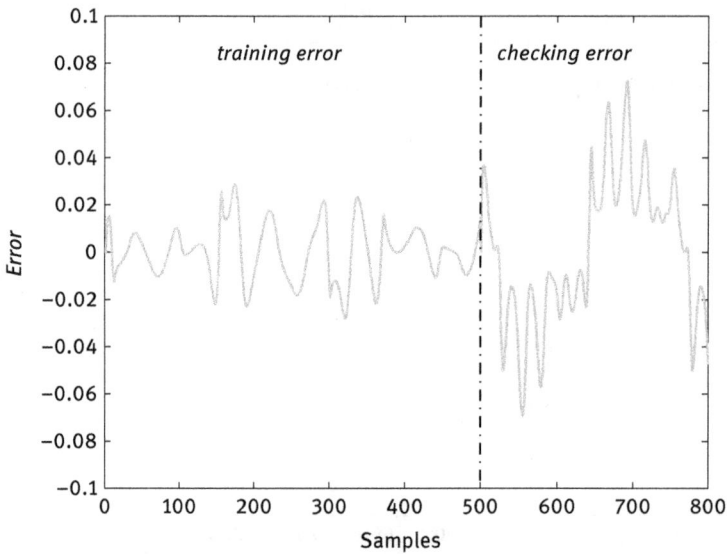

Fig. 10. Training and checking error using the square norm regularization.

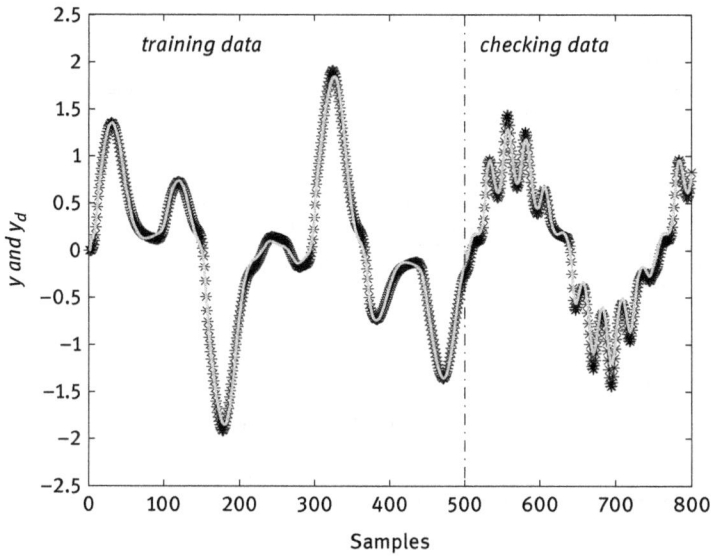

Fig. 11. The plant (dashed line) and its estimated version (continued line) using the adjusted parameters decay regularization.

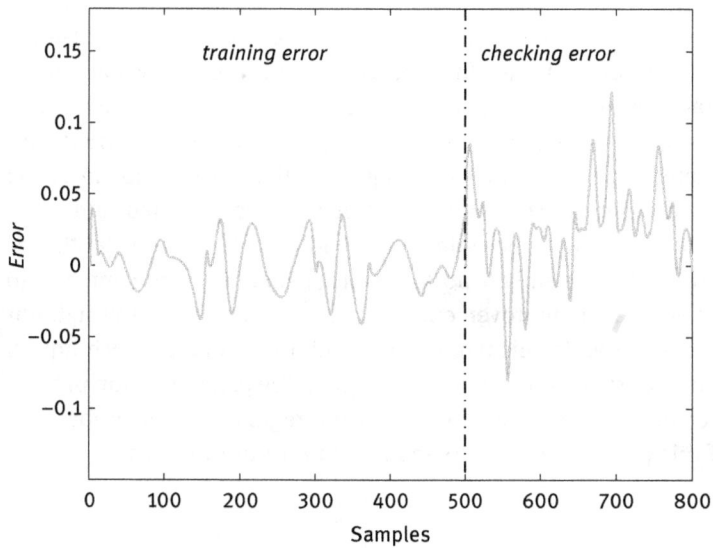

Fig. 12. Training and checking error using the adjusted parameters decay regularization.

Tab. 1. Performances of proposed regularization approaches.

Proposed approach		No regularization term	Square norm	Adjusted parameter decay
Example 1	TMSE	8.804×10^{-4}	3.196×10^{-4}	5.754×10^{-4}
	CMSE	0.078	0.0012	0.0035
Example 2	TMSE	9.096×10^{-4}	4.951×10^{-4}	6.51×10^{-4}
	CMSE	0.0278	0.0027	0.0069
Number of wavelets		12	12	12
Hyperparameters		$\lambda = 0$	$\lambda = 10^{-3}$	$\lambda_1 = 10^{-3}$ $\lambda_2 = 10^{-3}$ $\lambda_3 = 10^{-3}$ $\lambda_4 = 2 \times 10^{-3}$

The value of the hyperparameters has also an important influence on the whole training process, and so, on the generalization improvement.

7 Conclusion

Similar to neural networks, the generalization improvement of wavelet networks is considered to be a important issue. In fact, a given network may have good approximation accuracy, but could not perform well on unseen data. To deal with this kind of problem different regularization techniques have been carried out. Applied to wavelet networks, the treated techniques in this paper known as the square norm and the adjusted network parameters decay have provided numerous network parameters that could add more flexibility and control to the smoothness term. Consequently, this will, certainly, lead to a better generalization compared to a non regularized network. This improvement has been clearly shown through out the satisfactory results obtained in this paper using both regularization techniques. Another important parameter that has a serious impact on the generalization process is the hyperparameter that controls the strength of the regularization. In fact, the convenient choice of this parameter leads to an important future work that could be carried out.

Bibliography

[1]   M.Z.E.A. Skhiri and M. Chtourou. Wavelet Neural Networks Generalization Improvement. 10th *Int. Multi-Conf. on System, Signals and Devices*. March 2013.

[2]   H. Chen. On Regularized Negative Correlation Learning for Neural Network Ensembles. *IEEE Trans. on Neural Network,* 20(12):1962–1979, 2009.

[3]   J. Yaochu, T. Okabe and B. Sendhoff. Neural network regularization and ensembling using multi-objective evolutionary algorithms. *Congress on evolutionary computation*, Portland, USA, :1–8, 2004.

[4]   G. Ping, M.R. Lyu and C.L.P. Chen. Regularization parameter estimation for feedforward neural networks. *IEEE Trans on Networks Systems, Man and Cybernetics*, 33:35–44, 2003.

[5]   J. Larsen and L.K. Hansen. On optimal data split for generalization estimation and model selection. *IEEE Signal Processing Society Workshop on Neural Networks for Signal Processing IX*, 225–234, 1999.

[6]   J Larsen and L.K. Hansen. Generalization performance of regularized neural network models. *IEEE Workshop on Neural Networks for Signal Processing IV,* :42–51, 1994.

[7]   F. Girosi, M. Jones and T. Poggio. Regularization Theory and Neural Networks Architectures. *Neural Computation*, 7:219–269, 1995.

[8]   C. M. Bishop. *Neural Networks for Pattern Recognition*. Oxford University Press, Oxford, UK, 1995.

[9]   S. H. Chanz, H. W. Ngan and A. B. Rad. Improving Bayesian Regularization of ANN via Pre-training with Early-Stopping. *Neural Processing Letters*, 18(1):29–34, 2003.

[10] Y.Shao, G.N. Taff and S. J. Walsh. Comparison of Early Stopping Criteria for Neural Network-Based Subpixel Classification. *IEEE Geoscience and Remote Sensing Letters*, vol 8(1):113–117. 2011.

[11] G. Ramazan and Q. Min. Pricing and Hedging Derivative Securities with Neural Networks: Bayesian Regularization, Early Stopping, and Bagging. *IEEE Trans. on Neural Networks.*, 12(4), 2001.

[12] M.S. Iyer and R.R. Rhinehart. A novel method to stop neural network training. *American Control Conf.*, 2:929–933, Chicago, USA, 2000.

[13] L. Prechelt. Early stopping - but when, Neural Networks: Tricks of the Trade. *Springer-Verlag, Berlin Heidelberg vol. 1524 of LNCS.* chapter 2, pp. 55–69, 1998.

[14] L. Prechelt. Automatic early stopping using cross validation: quantifying the criteria. *Neural Networks*, 761–767, 1998.

[15] Y. Shao. Comparison of Early Stopping Criteria for Neural-Network-Based Subpixel Classification. *IEEE Geoscience and Remote Sensing Letters*, 8(1):113–117, 2011.

[16] C. S. Leung, H.J.Wang and J. Sum. On the Selection of Weight Decay Parameter for Faulty Networks. *IEEE Trans. on Neural Networks*, 21(8), 2010.

[17] Q. Zhang. On Analysis of weight decay regularisation in NNARX nonlinear identification. 5th *Int. Colloquium on Signal Processing and Its Applications*, CSPA 2009, 2009.

[18] M.H.F. Rahiman, M.N. Taib, R. Adnan and Y.M. Salleh. Analysis of weight decay regularization in NNARX nonlinear identification. 5th *Int. Colloquium on Signal Processing and Its Applications*, CSPA 2009, 2009.

[19] G. Gnecco and M. Sanguineti. The weight-decay technique in learning from data: An optimization point of view. *Computational Management Science*, 6(1), 2009.

[20] G. Gnecco and M. Sanguineti. Weight-decay regularization in reproducing Kernel Hilbert spaces by variable-basis schemes. *Computational Management Science*, 8:625–634, 2009.

[21] J. L. Bernier, J. Ortega, M. M. Rodriguez, I. Rojas and A. Prieto. An accurate measure for multilayer perceptron tolerance to weight deviations. *Neural Processing Letters*, 10(2):121–130, 1999.

[22] Z. Zhang. Learning algorithm of wavelet network based on sampling theory. *Neurocomputing*, 71:244–269, 2007.

[23] Z. Zhang. An Algorithm of Wavelet Network Learning from Noisy Data. *Intelligent Control and Automation*, :2746–2751, 2006.

[24] X. Gao and J. Zhang. A novel orthonormal wavelet network for function learning. *Int. Conf. on Advances in Natural Computation*, 2005.

[25] P. Zheng, W. Tang and J. Zhang. Generalization ability analysis of one-dimensional wavelet neural network by simulations. *Control and Decision Conf.*, :2506–2510, 2008.

[26] E. Ribes-Gomez, S. McLoone and G. Irwin. Orthogonal wavelet network construction using local regularisation. *First Int. IEEE Symp. on Intelligent Systems*, 1:271–276, 2002.

[27] A. Antoniadis and F. Jianqing. Regularization of Wavelet Approximations. *J. of the American Statistical Association*, 96(455):939–967, September, 2001.

[28] Q. Zhang and A. Benveniste. Wavelet networks. *IEEE Trans. on Neural Networks*. 3(6):889–898, 1992.

[29] Q. Zhang. *Wavelet networks: the radial structure and an efficient initialization procedure.* Technical Report of Link Ping University, LiTH-ISY-I-1423, 1992.

[30] Q. Zhang. Using wavelet network in nonparametric estimation. *IEEE Trans. on Neural Networks*, 8(2):227–236, 1997.

[31] B. Christophe, S. Mallat and J.J. Slotine. Wavelet Interpolation Networks. *European Symp. on Artificial Neural Networks*, ESANN'1998, Bruges, Belgium, :47–52, 1998.

[32] A. Rakotomamonjy and S. Canu. Frames, Reproducing Kernels, Regularization and Learning. *J. of Machine Learning Research.*, 6:1485–1515, 2005.

[33] F. Girosi. An Equivalence between Sparse Approximation and Support Vector Machines. *Neural Computation*, 10:1455–1480, 1998.

[34] L. Yu, K.K. Lai, S.Y. Wang and W. Huang. A Bias-Variance-Complexity Trade-Off Framework for Complex System Modeling. *Int. Conf. on Computational Science and its Application*, Berlin Heidelberg. :518–527, 2006.

[35] S. Geman, E. Bienenstock and R. Doursat. Neural networks and the bias/variance dilemma. *Neural Computations*, 4(1):1–58, 1992.

Biographies

Mohamed Zine El Abidine Skhiri born in Monastir (Tunisia) in 1959. He received the Bachelor of Science in Electrical Engineering from Syracuse University New York, USA in 1985, the Master of Science in Electrical Engineering from Syracuse University New York, USA in 1987 and the Doctorate in Electrical Engineering from the National school of Engineering of Sfax, Sfax, Tunisia on February 2013. He is currently a teaching assistant in the Electrical Engineering Department of the Institut Supérieur des Etudes Technologiques de Sousse , Sousse Tunisia. His current research interests include learning algorithms, Wavelet neural networks and their engineering applications.

Mohamed Chtourou was born in Sfax (Tunisia) in 1963. He received the Engineering Diploma in electrical engineering from the Ecole Nationale d'Ingénieurs de Sfax-Tunisia in 1989, the Diplôme d'Etudes Approfondies in Automatic Control from the Institut National des Sciences Appliquées de Toulouse-France in 1990, and the Doctorat in Process Engineering from the Institut National Polytechnique de Toulouse-France in 1993 and the Habilitation Universitaire in Automatic Control from the Ecole Nationale d'Ingénieurs de Sfax-Tunisia in 2002. He is currently a professor in the Department of Electrical Engineering of National School of Engineers of Sfax-University of Sfax-Tunisia. His current research interests include learning algorithms, artificial neural networks and their engineering applications, fuzzy systems, and intelligent control. He is author and co-author of more than forty papers in international journals and of more than sixty papers published in national and international conferences.

N. Bahri, A. Atig, R. Ben Abdennour, F. Druaux and D. Lefebvre

Multivariable Adaptive Neural Control Based on Multimodel Emulator for Nonlinear Square MIMO Systems

Abstract: This work describes multivariable adaptive neural control based on multimodel emulator for nonlinear square MIMO systems. Multimodel approach is an interesting alternative and a powerful tool for modelling and emulating complex processes. This paper deals with the identification of uncoupled nonlinear MIMO systems employing an uncoupled multimodel. Efficiency of this multimodel for systems emulation is illustrated in a multivariable adaptive neural control scheme. This emulator presents an important advantage in comparison to the classical emulators developed in the literature. Indeed, the multimodel emulator avoids the online adaptation procedure and the painful selection of the initialization parameter. The effectiveness of the proposed multimodel emulator and the control design for MIMO nonlinear systems are illustrated through numerical simulations.

Keywords: Uncoupled multimodel, indirect adaptive control, multivariable nonlinear systems, emulation, Neural Network.

1 Introduction

The presence of large nonlinearities in complex system dynamics makes the linearized models inefficient for controller design in many situations. Thus, nonlinear multivariable control has been the subject of constant research over several past decades. A promising alternative for the control of MIMO (Multi Input Multi Output) nonlinear dynamical systems is the implementation of neural networks [16, 20]. More precisely, there is an increasing interest for adaptive control methodologies based on neural networks [2, 3, 11, 21]. Since, the adaptive controller parameters are time varying, the online updating of these parameters is the key issue of adaptive control. However, most of this control schemes are usually based on algorithms that require a precise knowledge of the process dynamics and which need to evaluate the outputs variation against the inputs variation. In this context, an indirect adaptive control scheme based on recurrent neural networks controller (NC) and including a neural emulator (NE) has

N. Bahri, A. Atig, R. Ben Abdennour, F. Druaux and D. Lefebvre: N. Bahri[1], email:
bahri.nesrine@gmail.com, A. Atig[1], email: atigasmatica@yahoo.fr, R. Ben Abdennour[1], email:
Ridha.benabdennour@enig.rnu.tn, F. Druaux[2] email: fabrice.druaux@univ-lehavre.fr, and D.
Lefebvre[2] email: dimitri.lefebvre@univ-lehavre.fr, [1] Research Unit of Numerical Control of Industrial
Processes at ENIG (CONPRI), University of Gabes, Gabes Engineering School, Gabes, Tunisia,
[2] Electric and Automatic Engineering Research Group (GREAH), University of Le Havre, France

De Gruyter Oldenbourg, ASSD – Advances in Systems, Signals and Devices, Volume 5, 2018, pp. 165–182.
DOI 10.1515/9783110470468-010

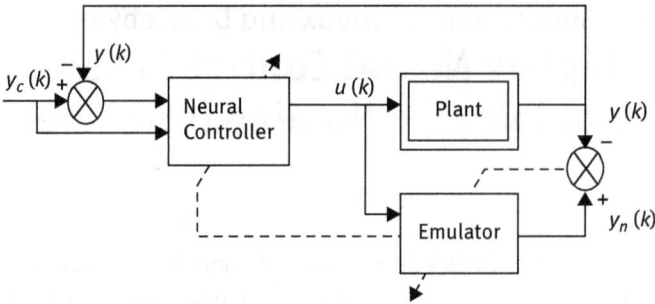

Fig. 1. Indirect neural control structure.

been previously proposed by the authors (Fig. 1) [1–5, 12, 21]. The proposed adaptation algorithm, used to update the NE and the NC parameters, is inspired from the real time recurrent learning (RTRL) algorithm.

This method drives the adaptation of all parameters whatever the process evolution. But, the neural emulator used in this scheme suffers from a strong limitation. Indeed, the obtained open and closed loop performance depends on the selection of a specific initialization parameter which is introduced to insure the starting of the parametric adaptation from zero initial conditions.

To overcome these problems, we proposed in our previous works [7, 8] to replace this neural emulator by a multimodel emulator for the adaptive control of SISO nonlinear systems. The contribution of this paper is to extend these works to square, uncoupled MIMO nonlinear systems. The multimodel approach is known as a powerful technique to overcome difficulties encountered in conventional representations and emulations of nonlinear systems. This approach is used in this paper to identify a set of models wich constitutes the multimodel emulator. This identification is made in open loop which reduces the computational complexity of the neural emulator and gets rid of problems caused by NE initialization parameter.

In this context, a multivariable adaptive neural control based on multimodel emulator is proposed for nonlinear square MIMO systems.

This paper is structured as follows. Section 2 describes the multimodel emulator for multivariable systems. A nonlinear MIMO system simulation illustrates the performance of the proposed ME in open loop emulation compared to the usual NE. In section 3, adaptive neural control based on multimodel emulation is developed. Simulation are proposed, illustrating the performance of the proposed ME in closed loop compared to the neural one. Conclusion and future works are enumerated in section 4.

2 Multimodel Emulation for Nonlinear MIMO Systems

2.1 Parametric Estimation Procedure

Our goal is to represent a multivariable nonlinear system with a decoupled multimodel. Let us define N_{IN} and N_{OUT}, respectively as the number of plant inputs and outputs where IN and OUT represent the set of inputs and outputs. In the following, we investigate square MIMO systems (ie. $N = N_{IN} = N_{OUT}$). In order to introduce the multimodel approach and for simplicity reasons any uncoupled MIMO model is decomposed into MISO (Multi Input Single Output) model. Each MISO model is modeled by a MISO multimodel: each MISO multimodel is attached to one of the system outputs. Thereafter identification tools available in [7, 8, 17–19], although developed in the context SISO are directly used to identify a MIMO system and to design multimodels for such systems. Without loss of generality, the multimodel is used up to now for nonlinear uncoupled multivariable systems.

The uncoupled structure, used for the lth MISO multimodel $l \in OUT$ (Fig. 2), is given by a state space representation in discrete time (k stands for the time variable)

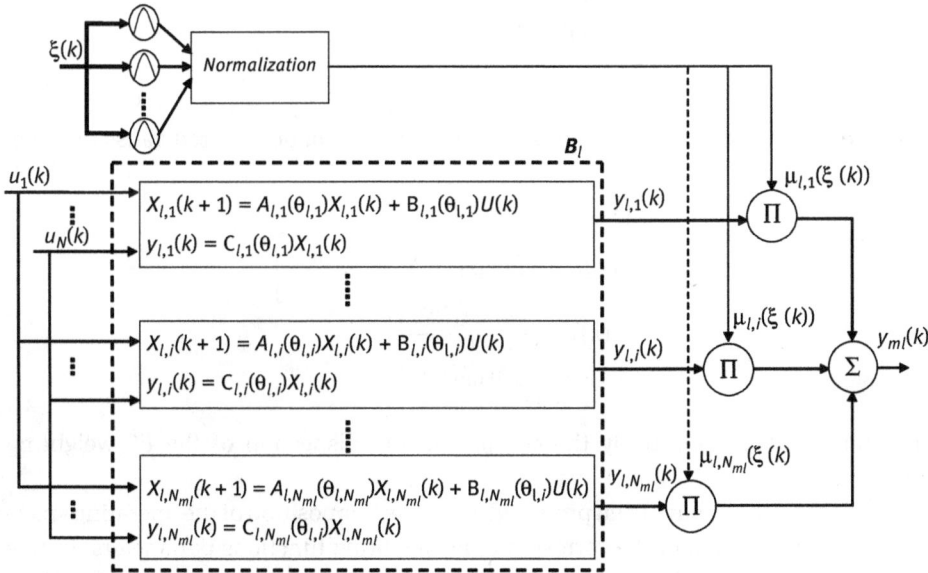

Fig. 2. ME base structure for multivariable nonlinear systems.

with sampling period ΔT:

$$X_{l,i}(k+1) = A_{l,i}(\theta_{l,i})X_{l,i}(k) + B_{l,i}(\theta_{l,i})U(k)$$
$$y_{l,i}(k) = C_{l,i}(\theta_{l,i})X_{l,i}(k) \tag{1}$$

where, $X_{l,i} \in \mathbb{R}^{n_{l,i}}$ and $U = [u_1(k), ..., u_l(k), ..., u_N(k)]^T \in \mathbb{R}^N$ are the state vector and the input vector, $y_{l,i}(k)$ and $n_{l,i}$ are the output and the dimension of the ith local model, respectively. $A_{l,i}(\theta_{l,i})$, $B_{l,i}(\theta_{l,i})$ and $C_{l,i}(\theta_{l,i})$ are, respectively, the state (or system) matrix, the input matrix and the output matrix of dimension $1 \times n_{l,i}$.

The multimodel representing the lth system output $y_{ml}(k)$ is defined by:

$$y_{ml}(k) = \sum_{i=1}^{N_{ml}} \mu_{l,i}(\xi(k))y_{l,i}(k) \tag{2}$$

N_{ml} is the number of local models for the lth system output.

$\xi(k) = [\xi_1(k), ..., \xi_l(k), ..., \xi_N(k)]^T$ is the decision variable vector which can be taken as the measurable state variables and/or input/output variables. In this work, the input vector is set as decision variables ($\xi(k) = U(k)$).

The local model contribution depends on the weighting function $\mu_{l,i}(\xi(k))$. A large choice of weighting functions is possible, they can be taken as triangular, sigmoidal, or gaussian functions and they must satisfy the following convex sums:

$$\sum_{i=1}^{N_{ml}} \mu_{l,i}(\xi(k)) = 1, \quad \forall i = 1...N_{ml},$$
$$0 \le \mu_{l,i}(\xi(k)) \le 1$$

Here the weighting functions $\mu_{l,i}(\xi(k))$ are obtained from normalized gaussian functions $w_{l,i}(\xi(k))$:

$$w_{l,i}(\xi(k)) = \prod_{j=1}^{N} \exp\left[-\frac{[\xi_j(k) - c_{l,ij}]^2}{\sigma_{l,ij}^2} \right]$$
$$\mu_{l,i}(\xi(k)) = \frac{w_{l,i}(\xi(k))}{\sum_{p=1}^{N_{m,l}} w_{l,p}(\xi(k))} \tag{3}$$

$c_{l,ij}$ and $\sigma_{l,ij}$ are respectively the centre and the dispersion of the ith weighting function.

Several methods have been proposed for the decomposition of the operating space and the determination of the corresponding weighting functions parameters. In this context, methods based on classification using Kohonen map and the Chiu classification were proposed for SISO nonlinear systems [10, 13]. The static characteristic of the considered system can also be used for the operating space decomposition [7, 8, 15, 17]. In the present work, the choice of N_{ml}, $c_{l,ij}$ and $\sigma_{l,ij}$ requires a priori informations about the static characteristic of the considered system.

Let us first define, for each MISO multimodel the vector of unknown parameters $\theta_{l,i}$ as follow:

$$\theta_l = [\theta_{l,1}^T \quad \cdots \quad \theta_{l,i}^T \cdots \quad \theta_{l,N_{ml}}^T]^T \tag{4}$$

where each column block $\theta_{l,i}$ is formed by the unknown parameters of the ith local model. The problem of parametric estimation arises in the following terms: using the input/output identification data we must find the vector $\hat{\theta}$ the estimate of θ under the minimum of a quadratic global criterion:

$$J_{ml} = \frac{1}{2} \sum_{k=1}^{N_H} (y_{ml}(k) - y_l(k))^2 \tag{5}$$

where, N_H is the number of training data.

Initially the Gauss–Newton's algorithm was used for the optimization procedure[18, 19]. In this work Levenberg–Marquardt's algorithm is used [14]:

$$\theta_l(it+1) = \theta_l(it) - \Delta_l(it)(H_l(\theta_l) + \lambda_l(it)I)^{-1}G_l(\theta_l) \tag{6}$$

$\theta_l(it)$ is the vector of the lth multimodel parameters at a particular iteration it and I is the identity matrix of appropriate dimension. At each iteration,. $H_l(\theta_l)$ and $G_l(\theta_l)$ are respectively the Hessian matrix and the gradient vector computed based on the calculation of sensitivity functions of output multimodel with respect to local models parameters:

$$H_l(\theta_l) = \frac{\partial^2 J_{ml}}{\partial \theta_l \partial \theta_l^T} \quad ; \quad G_l(\theta_l) = \frac{\partial J_{ml}}{\partial \theta_l} \tag{7}$$

$\Delta_l(it)$ is the relaxation coefficient introduced to minimises the criterion in the direction of vector $H_l^{-1}G_l$, $\lambda_l(it)$ is a regularization parameter. This scalar combines judiciously the gradient and the Gauss–Newton optimization methods to draw the profile of stability of the gradient algorithm and the speed of convergence of the Gauss–Newton method. The values of $\lambda_l(it)$ and $\Delta_l(it)$ are adjusted, usually by means of a heuristic based on the evolution of the criterion.

2.2 Numerical Example

2.2.1 Neural Emulator (NE) Limits

In this subsection, some simulation results are presented to demonstrate the problems of the classical neural emulator presented in the literature in [4]. To do this, consider the following two-input, two-output nonlinear system:

$$y_1(k) = \frac{1}{2}[(1.1 - 0.1z_{11}(k-1) - z_{21}(k-1))y_1(k-1)$$
$$+ z_{11}(k-1)u_1(k-1) + z_{21}(k-1)u_2(k-1)]$$

$$y_2(k) = \frac{1}{2}[(1.5 - z_{12}(k-1) - 0.2z_{22}(k-1))y_2(k-1)$$
$$+ z_{12}(k-1)u_1(k-1) + z_{22}(k-1)u_2(k-1)]$$

$$(8)$$

with:

$$z_{11}(k-1) = \frac{0.4 - 0.6y_1(k-1)}{1 + 0.2y_1(k-1)}; \ z_{21}(k-1) = \frac{0.6u_2(k-1)}{1 + 0.2y_1(k-1)}$$

$$z_{12}(k-1) = \frac{0.5u_1(k-1)}{1 + 0.5y_2(k-1)}; \ z_{22}(k-1) = \frac{0.7 - 0.07y_2(k-1)}{1 + 0.1y_2(k-1)}$$

As noted in the introduction and demonstrated in [6], the neural emulation performance depends on a single initialization parameter ε_e which is chosen arbitrarily to
ensure the starting of the system emulation with zero initial conditions of parameters.
In fact, an intuitive choice of the initialization parameter can affect the emulation
performance. For example, the choice of ε_e leads to 0.02 lead to the results depicted in
Fig. 3. Time is given in an arbitrary time unit. This figure confirms that the performance
is affected by the value of ε_e. Indeed, the neural emulator adapts itself to the variation
of the plant outputs with a relatively important output estimation errors. However,
another selection of $\varepsilon_e = 0.4$ allows the NE to provide a satisfactory estimation of the
precess outputs (Fig. 4).

In the next subsection we will demonstrate advantages of the multimodel
emulators.

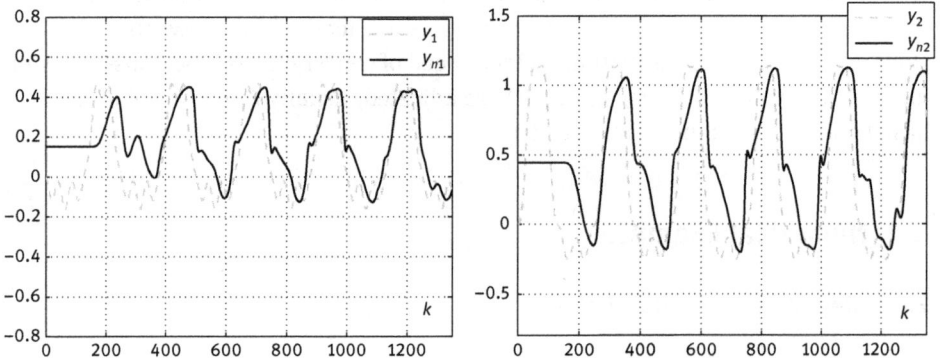

Fig. 3. Variation of real nonlinear system and Neural Emulator outputs ($\varepsilon_e = 0.02$).

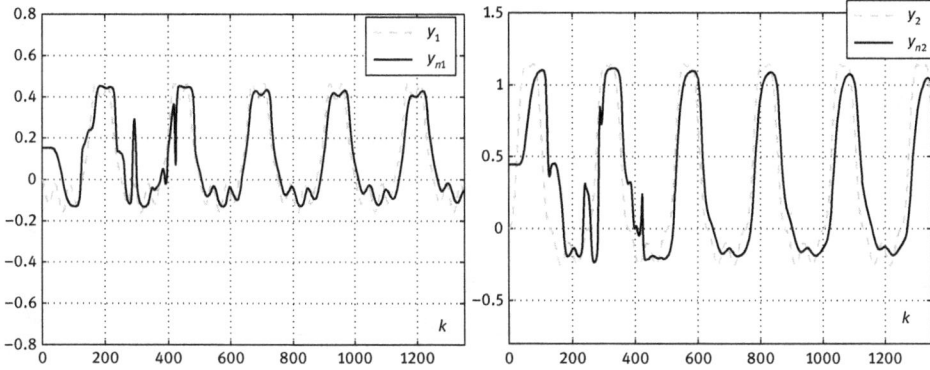

Fig. 4. Variation of real nonlinear system and Neural Emulator outputs ($\varepsilon_e = 0.4$).

2.2.2 Effectiveness of the multimodel emulator

In this subsection, a simulation study is presented to verify the effectiveness of the multimodel emulator for MIMO nonlinear systems. Consider the same nonlinear MIMO systems defined by (8).

The identification of the two MISO multimodels is realized with a global criterion (defined by (5)). The inputs $u_l(k)$, $l = 1, 2$ of the system, that will serve as decision variables for the weighting functions ($\xi_l(k) = u_l(k)$), consist of a signal with variable amplitude ($u_l(k) \in [-1, 1]$, $l = 1, 2$).

By exploiting the static characteristics of the system (Fig. 5), every MISO multimodel comprises ($N_m = 4$) sub-models and the weighting functions (Fig. 6) associated with each operating region are obtained by evaluating the expression (3) as follows:

$$y_{m1} \begin{cases} c_{1,11} = c_{1,21} = -1, \ c_{1,31} = c_{1,41} = 0.6, \ \sigma_{1,1} = 0.8 \\[2mm] c_{1,12} = c_{1,32} = -1.2, \ c_{1,22} = c_{1,42} = 1.4, \ \sigma_{1,2} = 0.6 \end{cases}$$

$$y_{m2} \begin{cases} c_{2,11} = c_{2,21} = -1.2, \ c_{2,31} = c_{2,41} = 1.4, \ \sigma_{2,1} = 0.6 \\[2mm] c_{2,12} = c_{2,32} = -0.8, \ c_{2,22} = c_{2,42} = 0.8, \ \sigma_{2,2} = 0.3 \end{cases}$$

We note that, in general, the local models must have a structure as simple as possible. Only the off-line validation phase of the base of models may increase the structure complexity.

In this case, a set of first order models provides sufficient precision. Indeed, the validation results of the Multimodel Emulation given in Fig. 7 show the variations of the real nonlinear system and the multimodel outputs ($y_l(k)$ and $y_{ml}(k)$, $l = 1, 2$).

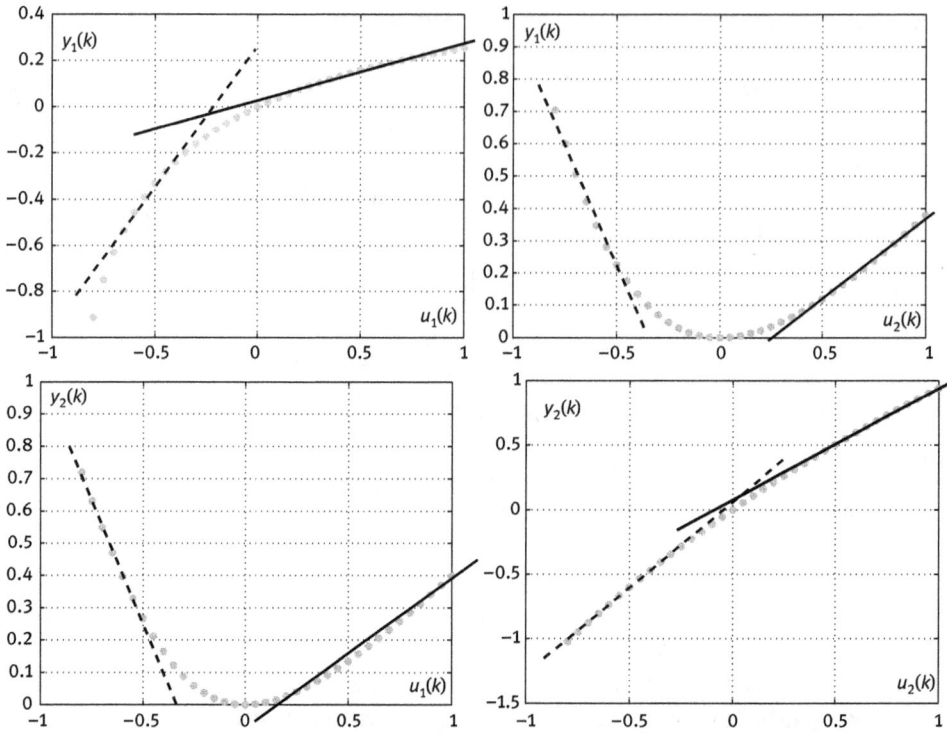

Fig. 5. The static characteristics of the considered real nonlinear system.

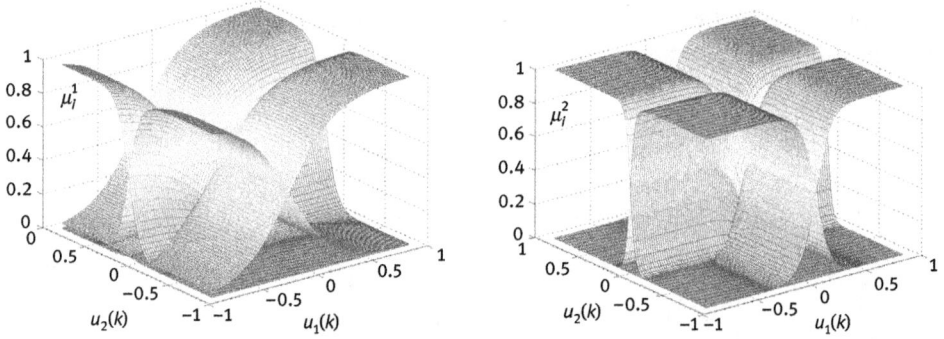

Fig. 6. Weighting functions.

The simulation results confirm that the uncoupled multimodel emulator offers a satisfactory modeling precision. These satisfactory obtained results are then compared to those obtained for neural emulation [2].

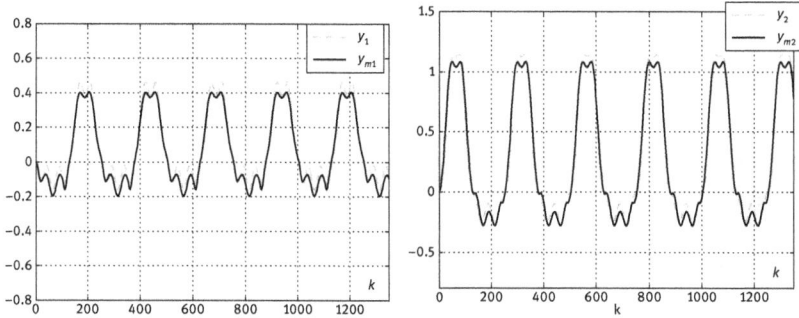

Fig. 7. Variation of real nonlinear system and Multimodel Emulator outputs.

To evaluate emulators performance, two indexes are used, for each outputs, such as the Mean square Error (MSE_l) and the Variance-Accounted-For (VAF_l) given by the following equations:

$$MSE_l = \frac{1}{N_H} \sum_{k=1}^{N_H} (y_{ml}(k) - y_l(k))^2 \tag{9}$$

$$VAF_l = \max\left\{1 - \frac{\text{var}\{y_{n_l}(k) - y_l(k) : k = 1...N_H\}}{\text{var}\{y_{n_l}(k) : k = 1...N_H\}}, 0\right\} \tag{10}$$

Table 1 summarizes these indexes calculated in open-loop emulation case for both neural and multimodel emulators. In Tab. 1, we note that neural emulation depend on the choice of the starting parameter ε_e. An intuitive and non systematic choice of the initialization parameter can affect the emulation performance. Indeed, using the neural emulator several tests were made to find the value of $\varepsilon_e = 0.4$ which gives relatively satisfactory emulation results. Results obtained using a multimodel emulator remains better than the ones obtained for neural emulator.

Tab. 1. MSE_l and VAF_l ($l = 1, 2$) for both outputs, in open-loop emulation case for both neural and multimodel emulators.

	Multimodel Emulator	Neural Emulator[4]
MSE_1	$1.4\,10^{-3}$	$2.51\,10^{-2}\,(\varepsilon_e = 0.02)$ $8.4\,10^{-3}\,(\varepsilon_e = 0.4)$
MSE_2	$3.2\,10^{-3}$	$1.32\,10^{-1}\,(\varepsilon_e = 0.02)$ $7.04\,10^{-2}\,(\varepsilon_e = 0.4)$
VAF_1	98.22%	$32.2\%\,(\varepsilon_e = 0.02)$ $80.74\%\,(\varepsilon_e = 0.4)$
VAF_2	98.82%	$29.25\%\,(\varepsilon_e = 0.02)$ $70.67\%\,(\varepsilon_e = 0.4)$

3 Adaptive Control Based on Multimodel Emulation for Nonlinear MIMO Systems

3.1 Multivariable Neural Control Strategy based on Multimodel Emulator

The neural controller (NC) is developed with fully connected recurrent neural networks. This structure is formed by $N_c = 2N$ neurons. The NC input signals are the N output error functions and the N desired outputs. These controller inputs are suitable both on tracking and regulation issues. The control inputs are the N first outputs of the network.

According to the dynamic activation of neurons, the controller outputs are calculated in discrete time by the following equations:

$$o_i(k) = e^{-|\tau_c(k-1)|\Delta T} o_i(k-1) \tag{11}$$
$$+ (1 - e^{-|\tau_c(k-1)|\Delta T}) D_i(k-1)$$

$$D_i(k) = \tanh \left(\sum_{j=1}^{N_c} \phi_{ij}(k) o_j(k) + z_i(k) \right)$$

o_i is the ith neuron state of controller, $u_i(k) = o_i(k)$, if $i \in 1, ..., N$ is the ith controller output, ϕ_{ij} is the controller weight from neuron j to neuron i and $1/|\tau_c|$ is the controller time parameter. $z_i(k) = y_{c_i}(k) - y_i(k)$ if $i \in \{1, ..., N\}$; $z_i(k) = y_{c_{i-N}}(k)$ if $i \in \{N+1, ..., 2N\}$.

Let us consider the instantaneous square error between desired outputs and measured outputs:

$$e_c(k) = \frac{1}{2} \sum_{l=1}^{N} (y_{c_l}(k) - y_l(k))^2 \tag{12}$$

The NC weights are updated by the minimisation of $e_c(k)$ according to an autonomous algorithm inspired from the RTRL:

$$\Delta \phi_{ij}(k) = |\eta_c(k)| \Delta T \sum_{l=1}^{N} (y_{c_l}(k-1) - y_l(k-1)) \frac{\partial y_l(k-1)}{\partial \phi_{ij}} \tag{13}$$

where, $\eta_c(k)$ is the NC adapting rate.

Using the multimodel emulator (2) the derivative of the output variation against NC weights $(\partial y_l(k)/\partial \phi_{ij}, l = 1, ..., N)$ is approximated by $(\partial y_{ml}(k)/\partial \phi_{ij}, l = 1, ..., N)$ according to equation (14):

$$
\begin{aligned}
\frac{\partial y_{ml}(k)}{\partial \phi_{ij}} &= \sum_{d=1}^{N} \frac{\partial y_{ml}(k)}{\partial o_d} \frac{\partial o_d(k)}{\partial \phi_{ij}} \\
&= \sum_{d=1}^{N} \frac{\partial y_{ml}(k)}{\partial o_d} Q_{dij}(k)
\end{aligned}
\tag{14}
$$

then the term $\dfrac{\partial y_{ml}(k)}{\partial u_d}$ can replace the term $\dfrac{\partial y_{ml}(k)}{\partial o_d}$:

$$
\frac{\partial y_{ml}(k)}{\partial u_d} = \sum_{i=1}^{N_{ml}} \mu_{l,i}(\xi(k)) \frac{\partial y_{l,i}(k)}{\partial u_d}
\tag{15}
$$

According to the NC dynamic behaviour given by (11), the sensitivity functions $Q_{dij}(k)$ are calculated as follows:

$$
\begin{aligned}
Q_{dij}(k) &= e^{-|\tau_c(k-1)|\Delta T} Q_{dij}(k-1) \\
&+ (1 - e^{-|\tau_c(k-1)|\Delta T}) \varphi_d(k-1) \psi_d(k-1)
\end{aligned}
\tag{16}
$$

For $d \in \{1, ..., N\}$, $z_d(k) = y_{c_d}(k) - y_d(k)$ so $\varphi_d(k)$ and $\psi_d(k)$ are computed as:

$$
\varphi_d(k) = \tan h' \left(\sum_{h=1}^{N_c} \phi_{dh}(k) o_h(k) + z_d(k) \right)
$$

$$
\psi_d(k) = \left(\delta_i^d o_j + \sum_{h=1}^{N_c} \phi_{dh}(k) Q_{hij}(k) - \frac{\partial y_{md}(k)}{\partial \phi_{ij}} \right)'
$$

For $d \in \{N+1, ..., 2N\}$, $z_d(k) = y_{c_d}(k)$ and $\varphi_d(k)$ and $\psi_d(k)$ are given :

$$
\varphi_d(k) = \tan h' \left(\sum_{h=1}^{N_c} \phi_{dh}(k) o_h(k) + z_d(k) \right)
$$

$$
\psi_d(k) = \left(\delta_i^d o_j + \sum_{h=1}^{N_c} \phi_{dh}(k) Q_{hij}(k) \right)
$$

Based on the same method as the one used for the weights matrix adaptation, an algorithm is defined in order to adapt the parameters $\eta_c(k)$ and $\tau_c(k)$:

$$
\Delta \eta_c(k) = \Delta T \sum_{l=1}^{N} (y_{c_l}(k-1) - y_l(k-1)) \frac{\partial y_{ml}(k-1)}{\partial \eta_c}
\tag{17}
$$

$$
\Delta \tau_c(k) = |\eta_c(k)| \Delta T \sum_{l=1}^{N} (y_{c_l}(k-1) - y_l(k-1)) \frac{\partial y_{ml}(k-1)}{\partial \tau_c}
\tag{18}
$$

with:

$$\frac{\partial y_{ml}(k)}{\partial \eta_c} = \sum_{d=1}^{N} \frac{\partial y_{ml}(k)}{\partial o_d} \frac{\partial o_d(k)}{\partial \eta_c}$$

$$\frac{\partial y_{ml}(k)}{\partial \tau_c} = \sum_{d=1}^{N} \frac{\partial y_{ml}(k)}{\partial o_d} \frac{\partial o_d(k)}{\partial \tau_c}$$

(19)

Let's define $\Xi_d^{\eta_c} = \partial o_d / \partial \eta_c$ and $\Xi_d^{\tau_c} = \partial o_d / \partial \tau_c$.

$\Xi_d^{\eta_c}$ and $\Xi_d^{\tau_c}$ are considered as small perturbations added to the dth neuron state consequently to small variations $\partial \eta_c$ and $\partial \tau_c$ of respectively η_c and τ_c [1–5]. For simplification, we set $\Xi_d^{\eta_c} = \Xi_d^{\tau_c} = \Xi_d$. Then we can define the functions Ξ_d as follows:

$$\Xi_d(k) = e^{-|\tau_c(k-1)|\Delta T} \Xi_d(k-1)$$
$$+(1 - e^{-|\tau_c(k-1)|\Delta T})\left(\varphi_d(k-1)\chi_d(k-1) + \frac{\varepsilon_c}{|\tau_c(k-1)|}\right)$$

(20)

where $\varepsilon_c = \frac{d\Xi_d}{dt}\Big|_{t=0} > 0$ guarantees the algorithm starting and accelerates the controller adaptation from zero initial conditions.

For $d \in \{1, ..., N\}$, $z_d(k) = yc_d(k) - y_d(k)$ then $\chi_d(k)$ is computed as:

$$\chi_d(k) = \sum_{h=1}^{N_c} \phi_h(k)\Xi_h(k) - \frac{\partial y_{md}(k)}{\partial \eta_c}$$

For $d \in \{N+1, ..., 2N\}$, $z_d(k) = yc_d(k)$ then:

$$\chi_d(k) = \sum_{h=1}^{N_c} \phi_h(k)\Xi_h(k)$$

Adaptive neural control is used primarily to ensure the stability of the closed loop system, the convergence of the control error to 0 and the rejection of the disturbances. For these reasons it is important to study the stability of the presented approach and its robustness with respect to perturbations. In this context, the stability of indirect adaptive control scheme has been proposed by the authors [5, 12, 21] for the neural emulator. First works [21] derived an on-line weights updating law from the Lyapunov approach. According to the neural controller adaptive learning rate parameter, sufficient conditions for stability were obtained. Then these works have been improved using Lyapunov stability and tracking errors dynamics [5]. Indeed, Lyapunov sufficient stability conditions for decoupled adaptive rates of the emulator and controller were determined. New adaptation strategies based on the tracking error dynamics and on the use of direct stability analysis were developed. In our future work we envisage enlarging the stability study for the proposed adaptive control based on a multimodel emulator and investigating the robustness of the approach with respect to perturbations.

3.2 Numerical Example

To illustrate the effectiveness of the adaptive neural controller based on multimodel emulator for nonlinear MIMO systems, we consider the same nonlinear process described by (8). The obtained performance will be compared with the NE ones. Both tracking and regulation problems are studied according to numerical simulations.

A first simulation includes for both outputs, a tracking phase [1, 2500] and also a regulation one [2500, 4500]. During the regulation phase, at $k = 3000$, disturbances (during 500 periods) in the form of a step input, with a magnitude of 10% of the control outputs, was injected to the system. The term ε_c is used to ensure the starting of the system with zero initial conditions of parameters. Using an arbitrary value of ε_c (for example $\varepsilon_c = 4$) we obtain the result illustrated in Fig. 8. This figure shows that good performances are obtained in tracking phase. A satisfactory perturbation rejection is, also, recorded in regulation phase.

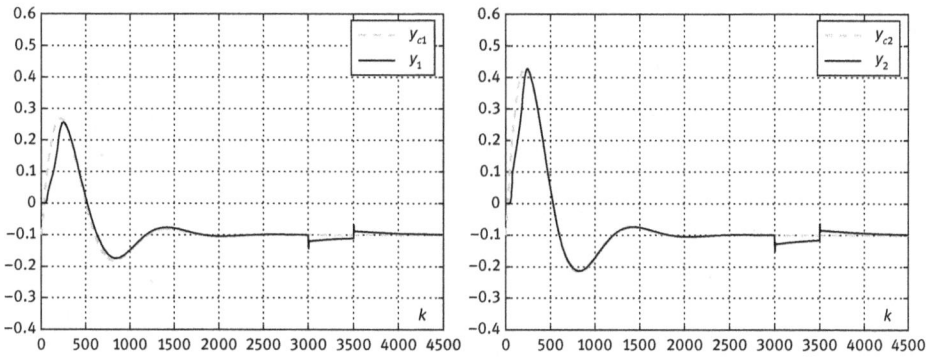

Fig. 8. Adaptive Neural Control based on Multimodel Emulator ($\varepsilon_c = 4$): desired and real system outputs.

On the other side, using a NE for adaptive neural control [4, 5] and the value of ε_e obtained previously in the phase of open loop case ($\varepsilon_e = 0.4$), we obtain the result illustrated in Fig. 9. For both system outputs, important oscillations are noted due to the updating algorithm. This figure shows that, in regulation phase, perturbations affect the control system performance and lead to high variance of the system outputs.

Table 2 summarizes the *MSE* calculated on adaptive neural control case based on neural and multimodel emulators. This table confirms that performance recorded using a multimodel emulator is far better than that using neural one [4, 5].

A second simulation with only a tracking phase is considered. In this case desired outputs speeds change at $k = 1500$. An arbitrary value of $\varepsilon_c = 10$ is taken. The simulation results are illustrated in Figs. 10 and 11.

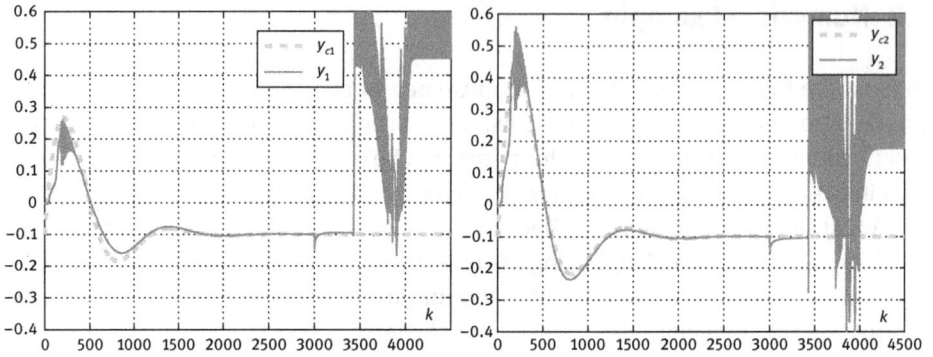

Fig. 9. Adaptive Neural Control based on neural emulator ($\varepsilon_e = 0.4$ and $\varepsilon_c = 4$): desired and real system outputs.

Tab. 2. *MSE* for both outputs, on Adaptive Neural Control case based on neural and multimodel emulators (tracking and regulation cases).

	Multimodel Emulator	Neural Emulator [4]
MSE_1	$4.55\,10^{-4}$	$8.66\,10^{-2}$
MSE_2	$7.32\,10^{-4}$	$7.24\,10^{-2}$

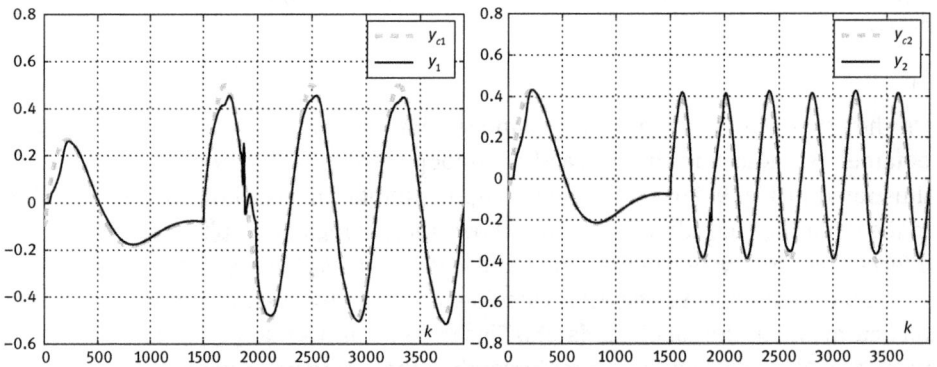

Fig. 10. Adaptive Neural Control based on Multimodel Emulator ($\varepsilon_c = 10$): desired and real system outputs.

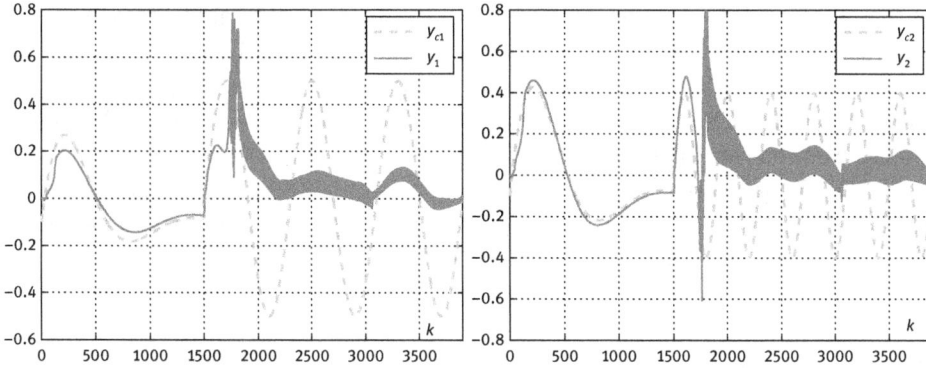

Fig. 11. Adaptive Neural Control based on neural emulator ($\varepsilon_e = 8$ and $\varepsilon_c = 10$): desired and real system outputs.

These figures show that results recorded in the case of the strategy advanced in this work is far better compared to the classical neural approach [4, 5]. In fact, there is no adaptation phase with the emulator multimodel so it is faster than the neural emulator.

Table 3 summarizes the *MSE* calculated on adaptive in this last control case. This results confirm our ascertainment.

Tab. 3. *MSE* for both outputs, on Adaptive Neural Control case based on neural and multimodel emulators (tracking case).

	Multimodel Emulator	Neural Emulator [4]
MSE_1	$1.6\,10^{-3}$	$6.89\,10^{-2}$
MSE_2	$1.3\,10^{-3}$	$5.7\,10^{-2}$

4 Conclusion

This article proposes an uncoupled multimodel emulator for multivariable indirect adaptive neural control scheme applied to nonlinear square and uncoupled MIMO systems. The proposed ME consists in finding a set of submodels with a simple linear structure and a set of appropriated weighting functions in order to combine these submodels to constitute the global model. This substitution reduces the computational complexity of the multivariable neural adaptive control scheme and avoids the problem related to the selection of the initialization NE parameter. The simulation results show clearly that the proposed multimodel emulator leads to good closed loop performance relatively to the case where classical neural emulator is applied. This results can be improved with an optimal choice of the local models structure (weighting functions structure). Whereas, if the considered system is unstable in open loop, the use of a stabilizing control to collect identification data is necessary.

In our future works, systematic generation of weighting functions structure for multimodel emulation will be envisaged. An extension of the multivariable neural adaptive control to non-square systems will also attract our interest.

Bibliography

[1]   A. Atig, F. Druaux, D. Lefebvre, K. Abderrahim and R. Ben Abdennour. A new neural adaptive control based on neural emulation of complex square systems. *Int. Review of Automatic Control, IREACO*, 3(6):612–623, 2010.

[2]   A. Atig, F. Druaux, D. Lefebvre, K. Abderrahim and R. Ben Abdennour. Neural emulator and controller with decoupled adaptive rates for nonlinear systems: application to chemical reactors. *Int. J. on Sciences and Techniques of Automatic Control and Computer Engineering*, (IJ-STA), 4(2):1298–1319, 2010.

[3]   A. Atig, F. Druaux, D. Lefebvre, K. Abderrahim and R. Ben Abdennour. Neural Emulation applied To Chemical Reactor. 7th *IEEE Int. Multi-Conf. on Systems, Signals and Devices* (SSD'10). Amman Jordan, 2010.

[4]   A. Atig, F. Druaux, D. Lefebvre, K. Abderrahim and R. Ben Abdennour. Neural Network Control for Large Scale Systems with Faults and Perturbations. 10th *IEEE Int. Conf. on Control and Fault-Tolerant Systems* (SysTol'10), Nice, France, 2010.

[5]   A. Atig, F. Druaux, D. Lefebvre, K. Abderrahim and R. Ben Abdennour. Adaptive control design using stability analysis and tracking errors dynamics for nonlinear square MIMO systems. *Engineering Applications of Artificial Intelligence* (EAAI), 25:1450–1459, 2012.

[6]   N. Bahri, A. Atig, R. Ben Abdennour, F. Druaux and D. Lefebvre. Emulation of Multivariable non square and nonlinear systems, 14th *Int. Conf. on Sciences and Techniques of Automatic control and computer engineering* (STA'13), Sousse, Tunisia, 2013.

[7]   N. Bahri, A. Messaoud and R. Ben Abdennour. A Multimodel Emulator For Non Linear System Controls. *Int. J. on Sciences and Techniques of Automatic Control and Computer engineering* (IJ-STA), 5(1):1500–1515, June 2011.

[8]   N. Bahri, A. Atig, R. Ben Abdennour, F. Druaux and D. Lefebvre. Multimodel and neural emulators for non-linear system: application to indirect adaptive neural control. *Int. J. of Modelling, Identification and Control* (IJMIC), 17(4):348–359, 2012.

[9]   N. Bahri, A. Atig, R. Ben Abdennour, F. Druaux and D. Lefebvre. Emulation of Multivariable non square and nonlinear systems, 14th International conference on Sciences and Techniques of Automatic control and computer engineering (STA'13), Sousse, Tunisia, 2013.

[10] S. L. Chiu. Fuzzy model identification based on cluster estimation. *J. of Intelligent and Fuzzy Systems*, 2:267–278, 1994.

[11] Y. Fu and T. Chai. Nonlinear multivariable adaptive control using multiple models and neural networks. *Automatica*, 43:1101–1110, 2007.

[12] E. Leclercq, F. Druaux, D. Lefebvre and S. Zerkaoui. Autonomous learning algorithm for fully connected recurrent networks. *Neurocomputing*, 63:25–44, 2005.

[13] M. Ltaief, A. Messaoud and R. Ben Abdennour. An optimal systematic determination of models' base for multimodel representation: Real time application. *Int. J. of Automation and Computing*, 11(6):644–652, 2014.

[14] Marquardt D., An algorithm for least-squares estimation of nonlinear parameters, *SIAM J. on Applied Mathematics*, 11(2):431–441, 1963.

[15] A. Messaoud, M. Ltaief and R.Ben Abdennour. Supervision based on a Multipredictor for an Uncoupled State Multimodel Predictive Control. 6th *Int. Conf. on Electrical Systems and Automatic Control*, JTEA'2010, Hammamet, Tunisia, 2010.

[16] K.S. Narendra and K. Parthasarathy. Identification and control of dynamical systems using neural networks. *IEEE Trans. on Neural Networks*, 1(1):4–27, 1990.

[17] R. Orjuela, D. Maquin and J. Ragot. Nonlinear system identification using uncoupled state multiple-model approach. *Workshop on Advanced Control and Diagnosis*, ACD'2006, Nancy, France, 2006.

[18] R. Orjuela, B. Marw, J. Ragot and D. Maquin. State estimation for non-linear systems using a decoupled multiple model, *Int. J. of Modelling, Identification and Control*, 4(1):59–67, 2008.

[19] R. Orjuela. *Contribution à l'estimation d'état et au diagnostic des systèmes représentés par des multimodèles*. PhD thesis, National Polytechnic Institute of Lorraine, Nancy-France, 2008.

[20] L. Tian, C. Collins. A dynamic recurrent neural network based controller for a rigid flexible manipulator system. Mechatronics, 14:3187–3202, 2004.

[21] S. Zerkaoui, F. Druaux, E. Leclercq and D. Lefebvre. Stable adaptive control with recurrent neural networks for square MIMO nonlinear systems. *Engineering Applications of Artificial Intelligence*, 12(4–5):702–717, 2009.

Biographies

Nesrine Bahri received her Engineering Diploma in Electric-Automatic engineering, in 2009, and the Master degree in Automatic Control and Intelligent Techniques, in 2010, from National School of Engineers of Gabes-Tunisia. Currently, she is pursuing her PhD thesis at CONPRI (Research Unit of Numerical Control of Industrial Processes at ENIG) and at GREAH (Electric and Automatic Engineering Research Group at Le Havre University). Her areas of interest include nonlinear process identification, multimodel and multicontrol approaches, neural and multimodel emulation and adaptive control.

Asma Atig received the Engineering Diploma in Electrical Engineering, in 2007, the Master degree in automatic control, in 2008, from the ENIS (National School of Engineering of Sfax-Tunisia.), the Ph.D. degree in Electrical Engineering from the ENIG (National School of Engineering of Gabes-Tunisia.) and from the University of Le Havre in Automatic, Signal Processing and Computing, in 2012. Actually she is a Teaching Assistant in Electrical Engineering Department at the High Institute of Industrial Systems of Gabes-Tunisia. She is member of CONPRI (Research Unit of Numerical Control of Industrial Processes at ENIG). Her areas of interest include nonlinear process identification, neural emulation and adaptive control.

Ridha Ben Abdennour received the Doctorat de spécialité degree from the Ecole Normale Supérieure de l'Enseignement Technique in 1987, and the Doctorat d'Etat degree from the Ecole Nationale d'Ingénieurs de Tunis in 1996. He is Professor in Automatic Control at the National School of Engineering of Gabes-Tunisia. He was chairman of the Electrical Engineering Department and the Director of the High Institute of Technological Studies of Gabes. He is the Head of the Research Unit of Numerical Control of Industrial Processes and is the Founder and honorary president of the Tunisian Association of Automatic and Numerisation. His research is on Identification, Multimodel & Multi-control approaches, Numerical Control and Supervision of Industrial Processes. He is the co-author of a book on Identification and Numerical Control of Industrial Processes and he is the author of more than 300 publications. He has participated in the organization of several Conferences and he was member of some scientific committees of congresses.

Fabrice Druaux received the B.S. degree in physic and mathematics in 1976 the M.S. in physic in 1981 and the Ph.D. degree in physic from University of Rouen (France) in 1986. Since 1988 he is an Asistant Professor at the Faculty of Sciences and Technology of Le Havre (France). Since 1999 he is with the G.R.E.A.H. (Electric and Automatic Engineering Research Group). His current research interests include modeling, control and fault detection using dynamical neural network. The principal applications are electro-technical processes such as motors and wind generators.

Dimitri Lefebvre is graduated from the Ecole Centrale of Lille (France) in 1992. He received the Ph.D. degree in Automatic Control and Computer Science from University of Sciences and Technologies, Lille in 1994, and the HAB. degree from University of Franche Comté, Belfort, France in 2000. Since 2001 he is Professor at Institue of Technology and Faculty of Sciences, University Le Havre, France. He is with the G.R.E.A.H. (Electric and Automatic Engineering Research Group). His current research interests include Petri Nets and DESs, learning processes, adaptive control, fault detection and diagnosis and applications to electrical engineering.

H. Huang, C. Gühmann and Y. Yu

Sliding Mode Based Engine Speed Control for an Automated Manual Transmission During Gear Shifting Process

Abstract: An effective engine speed control brings an obvious improvement to the automated manual transmission (AMT) shift quality during the gear shifting process. In this paper a second-order sliding mode control with the super twisting algorithm is applied. This control method has a good tracking performance and a strong robustness to the nonlinear uncertainty system. The results show that this advanced closed-loop control has huge advantages in comparison to conventional controls such as PID control and open control. The overshoot and steady-state error are separately reduced and prevented under all engine conditions, the AMT shift quality is improved.

Keywords: Sliding mode control, model-based control, engine speed control, super twisting algorithm, automated manual transmission, automotive.

1 Introduction

An automated manual transmission (AMT) is designed and improved on the basis of a manual transmission (MT). With the help of improved electronic technology and optimized control algorithms, AMT not only inherits the advantages from MT, such as lower weight, high efficiency and convenient maintenance, but also offers its own features, such as reduced life-cycle costs and enhanced low fuel consumption. These benefits make AMT widely used, especially in the electric vehicle [1, 2]. Since the AMT is shifted without load (Fig. 1), the power flow is interrupted during this process. It brings obviously vehicle speed reduction and a bad shift quality feeling to the driver when the shifting duration is longer, the uncontrolled engine speed also causes the difficulty for the subsequent clutch engagement control.

Based on the bus communication between the transmission control unit (TCU) and the engine control unit (ECU), a rapid and accurate engine speed control during the gear shifting process becomes a suitable method to reduce the AMT shift duration and improves the AMT shift quality. The goal is to synchronize the engine speed with the transmission main shaft speed during the gear shifting process. The challenge is to that the controller needs a rapid response and an accurate execution to the target speed request. Especially the controller should also have

H. Huang, C. Gühmann and Y. Yu: Institute Technische Universität Berlin, Germany,
email: hua.huang@campus.tu-berlin.de, email: yue.yu@mailbox.tu-berlin.de, email:
clemens.guehmann@tu-berlin.de, email: yue.yu@mailbox.tu-berlin.de

De Gruyter Oldenbourg, ASSD – Advances in Systems, Signals and Devices, Volume 5, 2018, pp. 183–202.
DOI 10.1515/9783110470468-011

a strong robustness against the engine dynamic changes such as engine load, environmental disturbance and system components aging. The conventional engine controls, such as proportional-integral-derivative (PID) control and open control using speed-torque-throttle three-dimensional map, are popular used as industrial controllers due to their simple realization and acceptable control results [3]. However, conventional engine controls cannot provide sufficient adaptability in all cases. As Fig. 2 shows, the conventional control has a relative large overshoot after the step response at 2 s, and when a load is added to the engine at 4.5 s, the controller cannot supply a sufficient controllability and causes a steady-state error. This causes an obvious engine speed oscillation during the gear shifting and it significantly decreases the AMT shift quality. These disadvantages make the follow-up shift quality optimization difficult, and cause huge manpower and financial costs.

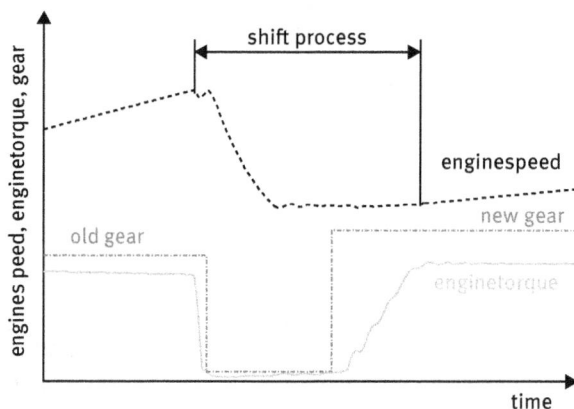

Fig. 1. Schematic diagram of the AMT upshift process.

The sliding mode control (SMC) is a variable structure control (VSC) method [4]. It is known as a robust control for uncertainty systems. The difference towards the conventional control is that its control law switches from one continuous structure to another in the state space instead of a continuous function of time. This brings the benefit that it has a good adaptation to the system disturbance and parameters changing. For the first-order sliding mode (FOSM), the control acts on the first time derivative of the sliding variable $\dot{s}$ ($\dot{s} = ds/dt$) to keep the system trajectories in the sliding set $s = 0$. Since this control law is based on the one degree-freedom control, it has disadvantages such as chattering from the high-frequency switching, which influences the system performance and even damages the actuators.

In order to overcome this disadvantage, some approaches are proposed. For example, a nonlinear reaching law containing an exponential term function of the sliding surface is introduced in [5], it reduces the control input signal chattering and also keeps high tracking performance in the steady-state regime. Also with the development of higher-order sliding mode (HOSM), this problem is solved and the control performances are improved [6, 7]. For the HOSM control (n-sliding, $n \geq 2$), it acts on higher time derivatives of the sliding variable, and drives the sliding variable to zero, i.e., $s = \dot{s} = \cdots = s^{n-1} = 0$. In [8] a second-order sliding mode control algorithm is applied to control an electronic throttle valve. The simulation and experiment results show that this HOSM has a better tracking performance than the PI control and FOSM control. The robustness of the HOSM controller is discussed in [9, 10], a diesel engine was used to investigate the speed control under different load changes.

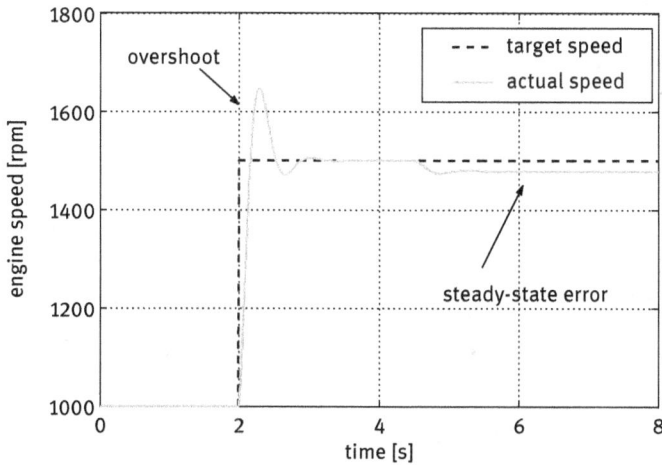

Fig. 2. Conventional control disadvantages.

In this paper, the second-order sliding mode controller with super twisting algorithm is chosen and applied to control the engine speed during the AMT gear shifting process in the model-based development. This second-order sliding mode control has the features that it does not require the time derivatives of the sliding variable and also effectively reduces the control signal chattering. This advantage makes it easy to replace the conventional controller and simply achieve the requirements. The controller parameters are easy to be determined and the tuning method is also described here. The tracking performance for the target speed and the robustness under different loads are also detailed investigated here.

The structure of this paper is as follows. Section 2 describes the control principle of the second-order sliding mode controller with the super twisting algorithm. Section 3 presents the simulation results for the tracking performance, robustness and the comparison with PI controller and open controller using speed-torque-throttle three-dimensional map. In section 4, this approved controller is applied to the AMT gear shifting process. Finally, a summary is concluded.

2 Second-Order Sliding Mode Control

The internal combustion engine of a passenger car is a nonlinear uncertainty system, the conventional controller does not provide enough adaptability in all conditions because of the dynamic changes, such as components aging and environmental disturbance. The second-order sliding mode controller has features such as high control accuracy and strong robustness, especially its control quality does not depend on the accuracy of the system model, so it is suitable to be applied for the engine speed control. The super twisting algorithm is chosen here, it has the advantage that it only needs the variable s for controlling (usually $\dot{s}$ is also needed for the second-order sliding mode control). And this algorithm is also possible to be used for the FOSM to weaken the chattering when the relative degree is 1: $\left(\dfrac{\partial}{\partial u} s = 0, \dfrac{\partial}{\partial u} \dot{s} \neq 0 \right)$.

2.1 Super Twisting Algorithm

The super twisting algorithm defines the control law for a system in (1) if the relative degree is 2: $\left(\dfrac{\partial}{\partial u} s = 0, \dfrac{\partial}{\partial u} \dot{s} = 0, \dfrac{\partial}{\partial u} \ddot{s} \neq 0 \right)$ or in (2) if the relative degree is 1: $\left(\dfrac{\partial}{\partial u} s = 0, \dfrac{\partial}{\partial u} \dot{s} \neq 0 \right)$ [11]:

$$\ddot{s} = \varphi(t, x) + \gamma(t, x) u \tag{1}$$
$$\ddot{s} = \varphi(t, x) + \gamma(t, x) \dot{u} \tag{2}$$

u is control input, s is sliding variable, x is a state vector, $|\varphi| \leq \Phi$, $0 < \Gamma_m \leq \gamma \leq \Gamma_M$. Φ, Γ_m, Γ_M are positive constants. The super twisting algorithm is expressed as follows (3, 4, 5) [11, 12]:

$$u(t) = u_1(t) + u_2(t) \tag{3}$$

$$\dot{u}_1 = \begin{cases} -u & |u| > 1 \\ -W \operatorname{sign}(s) & |u| \leq 1 \end{cases} \tag{4}$$

$$u_2 = \begin{cases} -\lambda |s_0|^\rho \operatorname{sign}(s) & |s| > s_0 \\ -\lambda |s|^\rho \operatorname{sign}(s) & |s| \leq s_0 \end{cases} \tag{5}$$

u_1 and u_2 are the control terms. W, λ and ρ are variable controller parameters. s is the sliding variable. s_0 is the boundary layer around sliding surface. Its corresponding sufficient conditions for finite time convergence are:

$$W > \frac{\Phi}{\Gamma_m}; \quad \lambda^2 \geq \frac{4\Phi\Gamma_M(W+\Phi)}{\Gamma_m{}^3(W-\Phi)}; \quad 0 < \rho \leq 0.5 \tag{6}$$

The convergence of the super twisting algorithm is shown in Fig. 3.

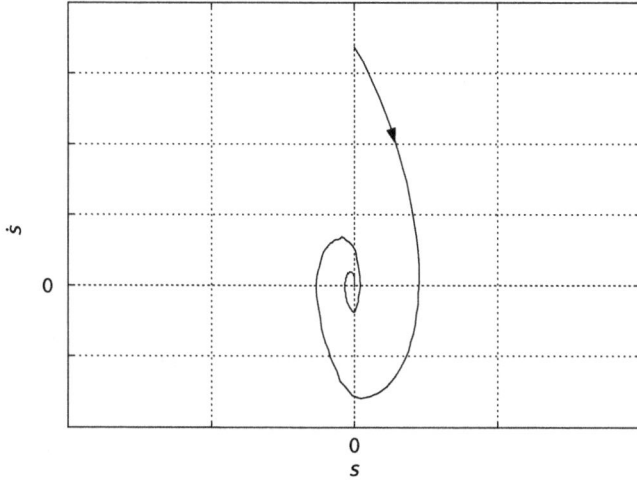

Fig. 3. Super twisting sliding mode phase plot.

2.2 Engine Model

In order to evaluate the second-order sliding mode speed tracking performance during the AMT gear shifting process, an engine model is used. A simplified engine model is expressed in (7):

$$\begin{cases} \dot{n}_e = \dfrac{30}{J_e\pi}[T_e(n_e,\theta) - T_{load}] \\ \dot{\theta} = -\dfrac{1}{\tau_\theta}\theta + \dfrac{1}{\tau_\theta}u \end{cases} \tag{7}$$

where n_e, T_e and J_e are the engine speed, torque and inertia. θ and u are the actual throttle position and the system control input (in the engine model u stands for the target throttle position). τ_θ is the time-delay constant. T_{load} denotes the engine load.

Set the engine speed n_e and θ as the state variable x ($x = (x_1, x_2)^T$), the target throttle position u as the control variable, then (7) can be transfered into a state-space (8):

$$\begin{cases} \dot{x}_1 = \sigma(x) + \delta \\ \dot{x}_2 = \alpha(x) + \beta u \end{cases} \tag{8}$$

where:

$$\sigma(x) = \frac{30}{J_e \pi} T_e(x_1, x_2), \; \delta = -\frac{30}{J_e \pi} T_{load}, \; \alpha(x) = -\frac{1}{\tau_\theta} x_2, \; \beta = \frac{1}{\tau_\theta}$$

The second differential of the engine speed n_e ($x_1 = n_e$) is expressed in (9) (based on the differential of the first state-space equation in (8) and δ is considered as constant):

$$\ddot{x}_1 = \phi(x) + \eta(x) u \tag{9}$$

where:

$$\eta(x) = \frac{30}{J_e \pi \tau_\theta} \frac{\partial T_e}{\partial x_2}, \; \phi(x) = \left(\frac{30}{J_e \pi}\right)^2 \frac{\partial T_e}{\partial x_1} T_e(x_1, x_2) + \frac{30}{J_e \pi} \frac{\partial T_e}{\partial x_1} \delta - \frac{30}{J_e \pi \tau_\theta} \frac{\partial T_e}{\partial x_2} x_2$$

2.3 Controller Design

Applied with super twisting algorithm in the engine speed control, the sliding variable defines as follows [13]:

$$s = e + c \dot{e} \tag{10}$$

$$e = n_{e\,actual} - n_{e\,target} \tag{11}$$

Then s first time derivative is given in the form:

$$\begin{aligned} \dot{s} = \dot{e} + c\ddot{e} &= \dot{x}_1 - \dot{x}_{1t} + c(\ddot{x}_1 - \ddot{x}_{1t}) \\ &= \sigma(x) + \delta - \dot{x}_{1t} + c(\phi(x) + \eta(x) u - \ddot{x}_{1t}) \\ &= \sigma(x) + c\phi(x) + \delta - \dot{x}_{1t} - c\ddot{x}_{1t} + c\eta(x) u \\ &= \varphi(t, x) + \gamma(t, x) u \end{aligned} \tag{12}$$

where:

$$x_1 = n_{e\,actual}, \; x_{1t} = n_{e\,target}, \; \gamma(t, x) = c\eta(x), \; \varphi(t, x) = \sigma(x) + c\phi(x) + \delta - \dot{x}_{1t} - c\ddot{x}_{1t}.$$

Simulation and experiment results show that the engine throttle valve has a better tracking performance when the relative degree is 1 than the relative degree is 2 [8], so c is chosen as a positive constant. Based on reference [11], (3), (4), (5) can be simplified when controlled systems are linear in control, u does not need to be bounded and

$s_0 = \infty$ (ρ is set to $1/2$ for the maximal possibility of second-order sliding realization):

$$u(t) = u_1(t) - \lambda|s|^{\frac{1}{2}} \operatorname{sign}(s) \tag{13}$$

$$\dot{u}_1 = -W \operatorname{sign}(s) \tag{14}$$

The corresponding control scheme is shown in Fig. 4. Since sliding variable s also depends on the differential $\dot{e}$, the super twisting based robust exact differentiator [14] is used.

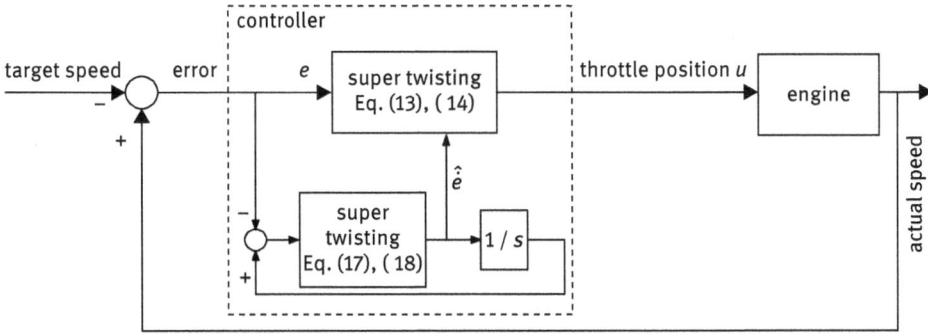

Fig. 4. Super twisting sliding mode control scheme.

2.4 Robust Exact Differentiator

The robust exact differentiator [14] is derived from the super twisting algorithm. Firstly consider an auxiliary equation:

$$\frac{d\hat{e}}{dt} = \hat{\dot{e}} \tag{15}$$

Assume a controller auxiliary equation output $\hat{e}$ perfectly tracks the computed speed error e:

$$\varepsilon := \hat{e} - e = 0 \tag{16}$$

Then $\hat{\dot{e}}$ equals to the time derivative of e. And $\hat{e}$, $\hat{\dot{e}}$ can be considered as the approximations of corresponding state variables e, $\dot{e}$. The super twisting algorithm is used here for the tracking task,

$$\hat{\dot{e}} = v - \vartheta|\varepsilon|^{\frac{1}{2}} \operatorname{sign}(\varepsilon) \tag{17}$$

$$\dot{v} = -\psi \operatorname{sign}(\varepsilon) \tag{18}$$

ψ and ϑ are positive constants, they are tuned empirically according to section 2.5.

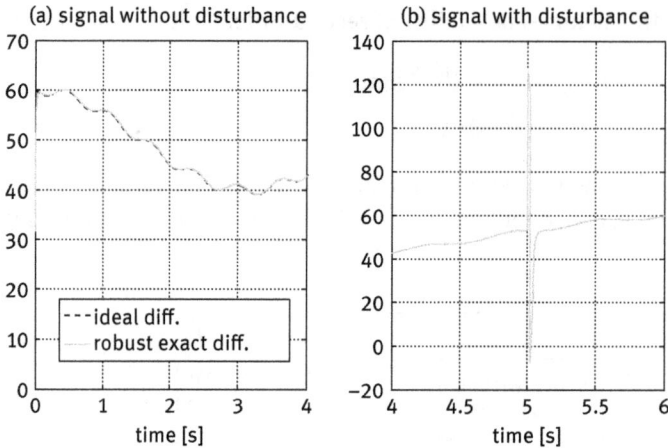

Fig. 5. Comparison of the robust exact differentiator with the ideal one. (a) left: without disturbance, (b) right: with disturbance.

Figure 5(a) shows the comparison results of the robust exact differentiator with the ideal one (du/dt) to a continuous signal $(f(t) = 50t + 10 \sin(t) + 0.1 \cos(10t))$, Fig. 5(b) is the robust exact differentiator performance when a disturbance (a pulse signal, amplitude is 5, period is 10 s, pulse width is 0.2 %) is added into this signal at 5 s. It can be proved that the robust exact differentiator has the similar behavior as the ideal one, and it also has a good robustness against the disturbance.

2.5 Controller Parameter Tuning Method

The tuning of the second-order sliding mode control parameters W and λ (see in (13), (14)) are similar to the Ziegler–Nichols rules, first W then λ. The W influences the oscillation in the steady state and λ influences the response speed. Figure 6 shows the tuning method of the W and λ variations to a step signal response. Firstly, parameter λ is set to zero, W is increased from zero till the engine speed begins to oscillate, as shown in Fig. 6(a). W is chosen between 80 % and 100 %, the detailed value choice depends on the speed oscillation in the steady state. Then λ is tuned as the same step. With the increasing of λ value, the system speed response time decreases, see in Fig. 6(b). The detailed λ valve is chosen based on the compromise of the system response time and the overshoot because when the response time decreases the overshoot increases.

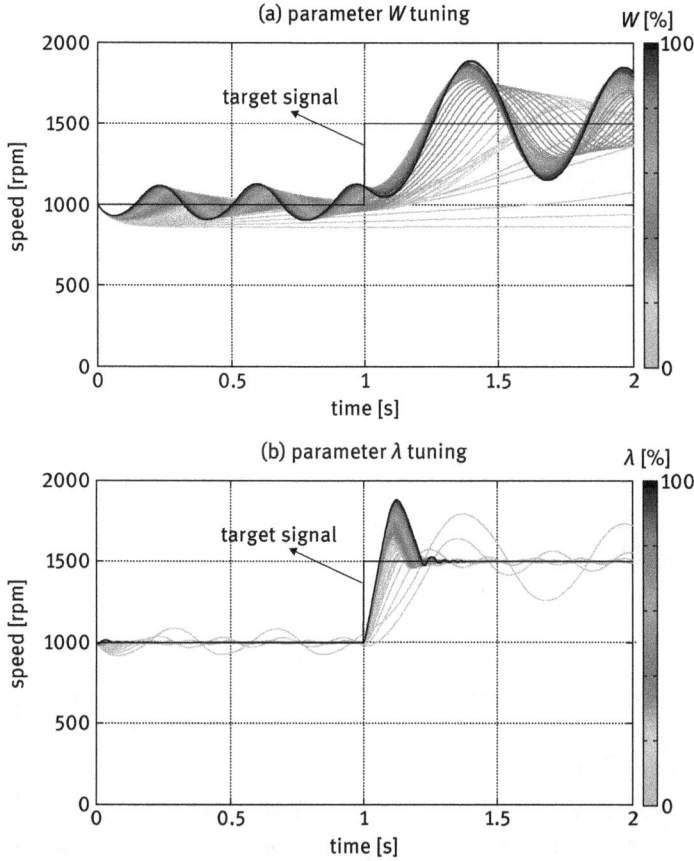

Fig. 6. Response performance in different W and λ. (a) top: W tuning, (b) bottom: λ tuning.

3 Testing and Comparison

In this section, the second-order sliding mode controller is tested in the following content. Firstly, the controller real-time capability is verified based on a throttle valve through the rapid control prototyping (RCP). Then, the controller convergence, tracking performance and robustness are tested in the engine model and the simulation results are also compared with the conventional controllers.

3.1 Real-Time Capability

The torque of the gasoline internal combustion engine is regulated by the position of the throttle valve, shown in Fig. 7, and it is also the main control object by the second-order sliding mode controller. Before the sliding mode control is tested in the engine system, it is necessary to verify it in the throttle valve system and compared with other controllers.

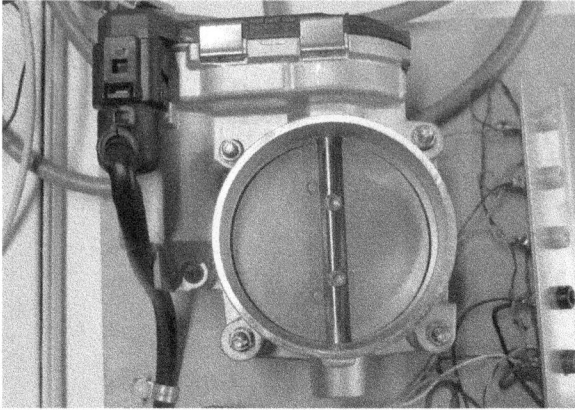

Fig. 7. Throttle valve.

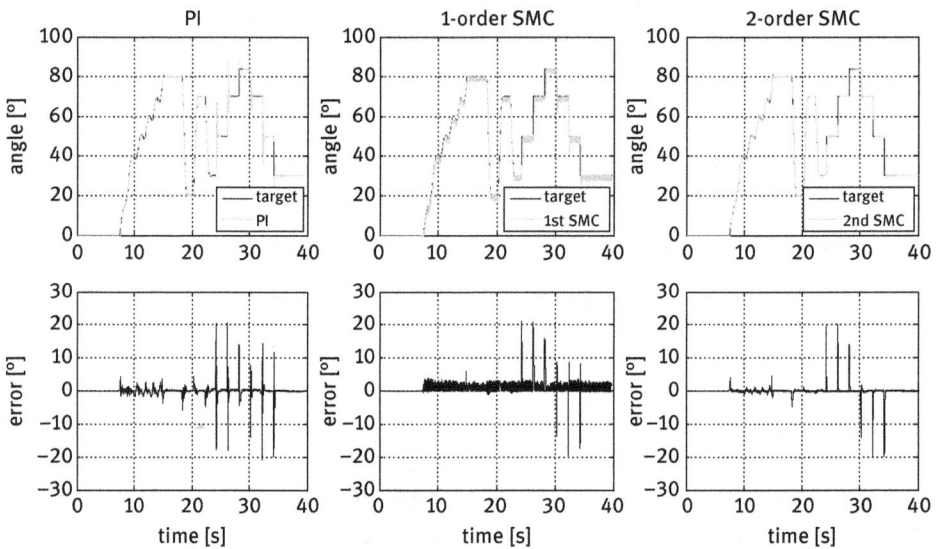

Fig. 8. Comparison results of different controller tracking performance in the throttle valve. (a) left: PI control, (b) middle: first-order SMC, (c) right: second-order SMC.

Here the controllers are implemented on a dSPACE® MicroAutoBox with a sampling period of $T_s = 1$ ms. Firstly, continuous signals with different angle changes are used to test the controller friction overcome ability and steady-state behavior, then a discontinuous signal is used to test the controller response behavior. PID control, first-order and second-order sliding mode control are used here to track the target angle signal. Since the derivative term in the PID controller is sensitive to the environmental noise, here only proportional and integral ones are used. Figure 8 shows the tracking performance results. From the comparison, the first-order sliding mode control has a smaller overshoot and better stable control results than PI control, but its chattering is obvious as described before. Second-order sliding mode control can notably suppress it and has a good tracking performance at the same time. Additionally, it can also state that the sliding mode controller is real-time capable. Since the chattering problem, which gives rise to the mechanical components damage and induces uncomfortable noise, the first-order sliding mode is unpractical to the real system, here the attention is only paid on the comparison of the conventional control and second-order sliding mode control.

Following the verification in the throttle valve system, the second-order sliding mode control is then tested with the engine plant in the the model-in-the-loop (MiL) simulation. The test content includes convergence, speed response and robustness.

3.2 Finite-Time Convergence

Firstly, the second-order sliding mode controller finite-time convergence is verified. Figure 9 respectively shows the engine speed responses in the s and $\dot{s}$ phase when a step signal is given from 1000 rpm to 1500 rpm and then the engine is disturbed with 40 Nm load. The sliding variables s and $\dot{s}$ converge to zero after a finite time duration.

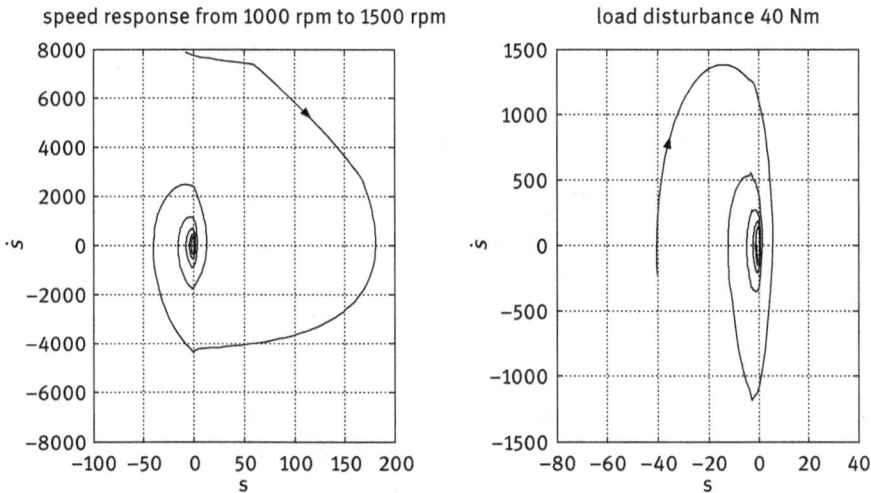

Fig. 9. Super twisting sliding mode convergence simulation. (a) left: speed response, (b) right: load disturbance.

3.3 Tracking Performance

Figure 10 shows the controllers (second-order sliding mode control, PI control and open control) tracking performance under a continuous signal. The second-order sliding mode control shows a better tracking performance than the conventional controls.

During the AMT gear shifting process, the engine speed value is requested to step to a new one based on the shifted transmission gear ratio, so the tracking performance for a step signal is also compared. The comparison results are shown in Fig. 11. The open control, which uses speed-torque-throttle three-dimensional map to calculate the desired throttle valve, has a smooth but relative slow response, it may be suitable for the engine speed control during the AMT gear shifting process since the speed control duration can be longer (depends on the gear shifting time and mechanical constraints) and a better clutch control can also reduce the speed difference during the engagement, but this method does not fit for the rapid and accurate response requirements here, so this control method is no longer discussed. Table 1 lists the tracking performance indexes of these two controllers. e_{steady} and $e_{overshoot}$ are calculated through (19) and (20). In order to compare the step response in the whole ranges (from 1000 rpm to 4000 rpm), n_{estep} and $n_{esteady}$ (in the denominator of the equations) are set to 1000 rpm. t_d is the delay time when the engine speed first reaches the 50 % of the steady-state value.

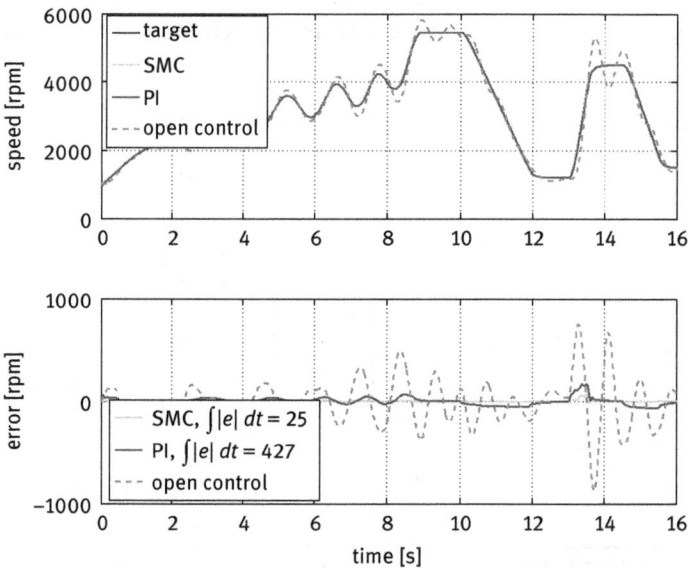

Fig. 10. Simulation results of continuous signal tracking performance.

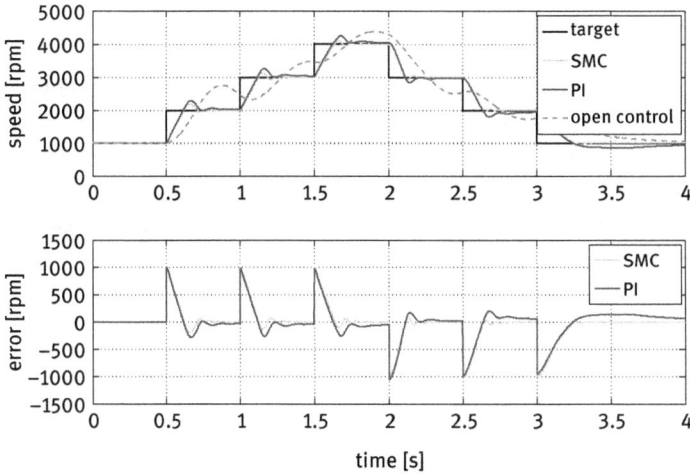

Fig. 11. Simulation results of step response.

Tab. 1. Tracking error for step signal.

speed [rpm]	controller	t_d [ms]	e_{steady} [%]	$e_{overshoot}$ [%]
1000–2000	PI	62	−2.40	28.33
	SM	66	0.14	18.97
2000–3000	PI	60	−3.27	26.27
	SM	66	0.16	12.73
3000–4000	PI	62	−4.74	25.80
	SM	70	0.30	10.54
4000–3000	PI	70	2.08	16.87
	SM	66	0.01	8.86
3000–2000	PI	78	5.50	20.11
	SM	80	−0.18	4.70
2000–1000	PI	107	13.82	13.82
	SM	113	−0.19	1.51

$$e_{steady} = \frac{n_{esteady} - n_{etarget}}{n_{estep}} \cdot 100\% \qquad (19)$$

$$e_{overshoot} = \frac{n_{eovershoot} - n_{esteady}}{n_{esteady}} \cdot 100\% \qquad (20)$$

3.4 Robustness

The robustness of the engine speed is an important evaluation index in engine control. Figure 12 shows the engine speed responses at various load conditions. From this comparison, the second-order sliding mode control shows a better robustness than the PI control. The comparison results are listed in Tab. 2. e_{RMS} shows the root-mean-square error between target speed and actual speed. $\int |e| \, dt$ describes the error integral during the engine speed response in every 0.5 s.

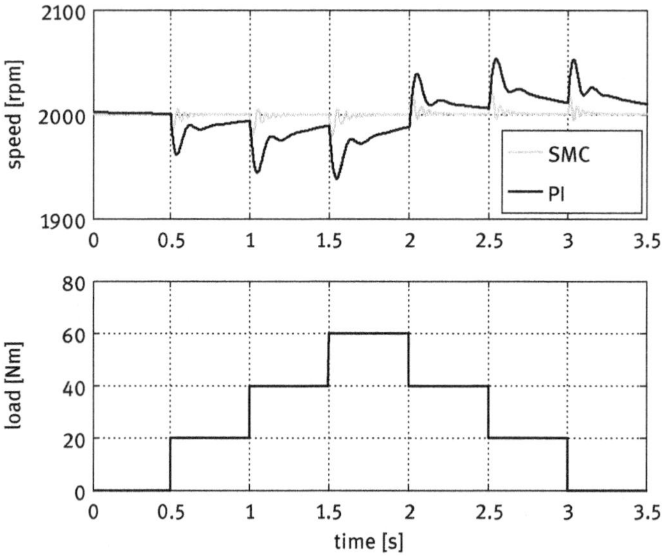

Fig. 12. Simulation results of robustness under different load.

4 AMT Gear Shifting Application

Based on the above discussion, the second-order sliding mode control shows a better control effect than the conventional one. The next step is to apply this control algorithm to the AMT gear shifting process. The schematic diagram of MiL simulation is shown in Fig. 13. The AMT equipped vehicle is modeled with Modelica® [15, 16], the corresponding TCU, ECU control algorithms and driver's command are developed with MATLAB®/Simulink®. In order to integrate both tools and to achieve the MiL simulation in PC, Silver® is used [17]. It co-simulates the Modelica® based model in FMU (Functional Mockup Units) file and MATLAB®/Simulink® based controller in dll (dynamic link library) file.

Tab. 2. Robustness under different load.

load [Nm]	controller	$\int|e|dt$	e_{RMS} [rpm]	e_{max} [%]
0–20	PI	6.72	8.16	1.90
	SM	0.69	2.86	0.70
20–40	PI	11.51	11.82	2.79
	SM	1.04	4.10	0.97
40–60	PI	12.94	13.02	3.07
	SM	1.13	4.38	1.02
60–40	PI	6.26	8.72	1.93
	SM	0.93	3.77	0.90
40–20	PI	11.15	10.95	2.66
	SM	0.95	3.86	0.93
20–0	PI	10.09	10.20	2.63
	SM	0.76	3.16	0.82

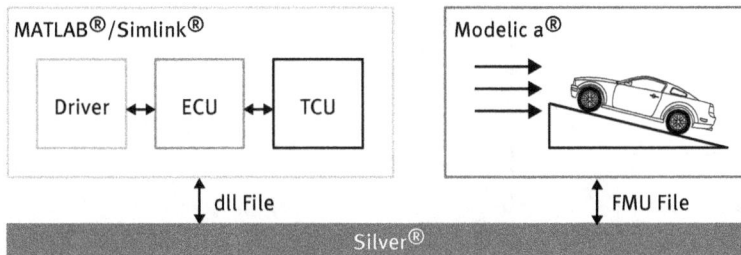

Fig. 13. Schematic diagram of MiL process.

The AMT gear shifting procedure includes the three phases: clutch opening, gear shifting and clutch engagement.

1. In the clutch opening phase, the engine load reduces to zero, the engine speed increases suddenly and even causes a "grating" noise when the accelerator pedal is too deep.
2. In the gear shifting phase the transmission input speed value is changed to a new one, the speed difference between the transmission main shaft and the engine crank shaft occurs.
3. In the clutch engagement phase, the clutch load is added into the engine again.

The engine speed intervention takes an important role in these phases, it keeps the engine speed at a constant value when the clutch is opened and synchronizes the engine speed with the transmission main shaft speed during the gear shifting phase. In the clutch engagement phase, the engine torque control is traditionally applied to prevent the engine speed from decreasing because of the load. With the introduced sliding mode control, the engine torque control is needless while the sliding mode controller has a good robustness to the load, as shown in Fig. 12. Finally, after the clutch is engaged, the controller calculates a corresponding time for the intervened throttle position back to the drivers desired position. The flowchart of this process is expressed in Fig. 14.

The gear shifting duration lasts approximately 0.5 s in the model-based AMT vehicle, it is longer than the designed response time, so the control parameters of the second-order sliding mode are intentionally slightly modified. Figure 15 shows the AMT up shifting from 2nd to 3rd and down shifting from 3rd to 2nd in a flat road.

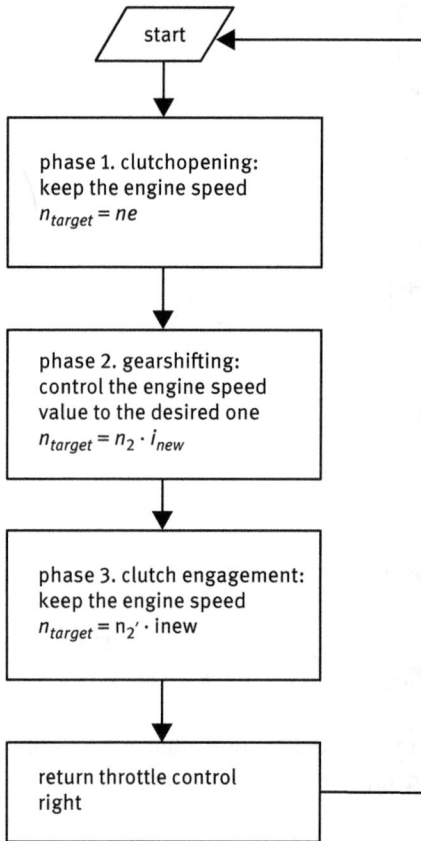

Fig. 14. Flowchart for the AMT gear shifting speed control.

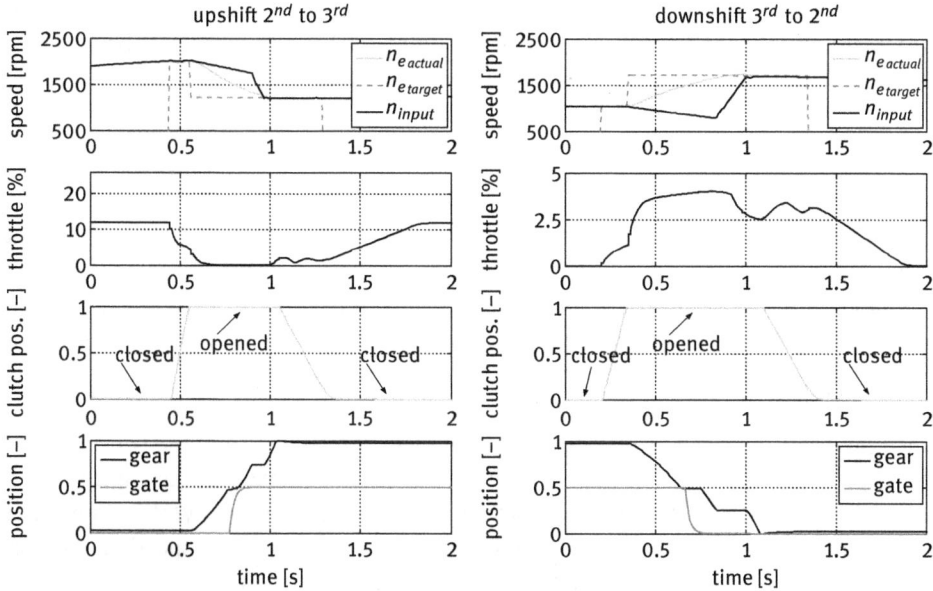

Fig. 15. Simulation results of gear shift between 2nd and 3rd. (a) left: upshift, (b) right: downshift.

Tab. 3. Engine speed control for different shiftings.

process	t [s]	e_{steady} [%]	$e_{overshoot}$ [%]
$1^{st}-2^{nd}$	0.51	−0.26	−0.77
$2^{nd}-3^{rd}$	0.45	0.65	−0.82
$3^{rd}-4^{th}$	0.38	−0.34	−1.00
$4^{th}-5^{th}$	0.32	0.62	−0.97
$5^{th}-4^{th}$	0.24	0.25	1.59
$4^{th}-3^{rd}$	0.37	0.22	1.37
$3^{rd}-2^{nd}$	0.54	0.17	0.87
$2^{nd}-1^{st}$	0.32	0.32	1.32

From the simulation results, it can be found that the engine speed almost keeps a constant value when the clutch opens (at about 0.45 s in Fig. 15(a) and about 0.36 s in Fig. 15(b)), the speed regulation completes before the clutch begins to engage, and while the clutch engages, the engine speed also keeps a constant value. Table 3 shows the detailed information of the engine speed control for different gear shifting, t denotes the time when the engine speed $n_{eactual}$ reaches the target value $n_{etarget}$, e_{steady} is the steady-state error and $e_{overshoot}$ is the overshoot error. It shows that the sliding mode controller effectively regulates the engine speed in different gear shifting processes and that the steady-state error and the overshoot error are satisfying.

5 Conclusion

In this paper, a second-order sliding mode controller with the super twisting algorithm is applied to an AMT system during the gear shifting process. This control method has the same input and output interface as the conventional control such as PID control and open control, it makes controller replacement become easy. Its implementation is also simple, the tuning of the control parameters are similar to the Ziegler-Nichols rules. Simulation results show that this control method has a better tracking performance (more rapidly and accurately) and a stronger robustness than the conventional control. This brings a big benefit to the shift quality optimization during the AMT gear shifting process. Combined with an appropriate control strategy, MiL simulation results show that this controller can effectively reduce the engine speed oscillation and regulate the engine speed within the gear shifting duration.

Bibliography

[1] G. Xiong, J. Xi, Y. Zhai, Y. Hu, Y. Yu and H. Chen. Development of pneumatically automatic mechanical transmission for a pure electric garbage truck. *IEEE Int. Conf. on Industrial Technology*, (ICIT), :1108–1112, Vi a del Mar, March, 2010.
[2] C. Liao, J. Zhang, and H. Zhu. A study of shift control algorithm without clutch operation for automated manual transmission in the parallel hybrid electric vehicle. *World Automotive Congress*, FISITA, Barcelona, May 2004.
[3] Zaimin Zhong, Qiang Lv and Guoling Kong. Engine speed control for the automatic manual transmission during shift process. 2nd *Int. Conf. on Consumer Electronics, Communications and Networks* (CECNet), :1014–1017, Yichang, April, 2012.
[4] R. A. DeCarlo, S. H. Zak, and G. P. Matthews. Variable structure control of nonlinear multivariable systems: a tutorial. *Proceedings of the IEEE*, 76(3):212–232, 1988.
[5] C. J. Fallaha, M. Saad, H. Y. Kanaan and K. Al-Haddad. Sliding-mode robot control with exponential reaching law. *IEEE Trans. on Industrial Electronics*, 58(2):600–610, 2011.
[6] T. Floquet, J.-P. Barbot and W. Perruquetti. Higher-order sliding mode stabilization for a class of nonholonomic perturbed systems. *Automatica*, 39(6):1077–1083, 2003.
[7] G. Bartolini, A. Ferrara, E. Usai and V.I. Utkin. On multi-input chattering-free second-order sliding mode control. *IEEE Trans. on Automatic Control*, 45(9):1711–1717, 2000.
[8] M. Reichhartinger and M. Horn. Application of higher order sliding-mode concepts to a throttle actuator for gasoline engines. *IEEE Trans. on Industrial Electronics*, 56(9):3322–3329, 2009.
[9] M. Khalid Khan, Keng Boon Goh, and Sarah K. Spurgeon. Second order sliding mode control of a diesel engine. *Asian J. of Control*, 5(4):614–619, 2003.
[10] K. B. Goh, S. K. Spurgeon, and N. B. Jones. Higher-order sliding mode control of a diesel generator set. *Proc. of the Institution of Mechanical Engineers, Part I: J. of Systems and Control Engineering*, 217(3):229–241, 2003.

[11] W. Perruquetti and J. P. Barbot. *Sliding Mode Control in Engineering*. Marcel Dekker, Inc., New York, U.S.A., 2002.

[12] A. Levant. Sliding order and sliding accuracy in sliding mode control. *Int. J. of Control*, 58(6):1247–1263, 1993.

[13] J.-J. E. Slotine and W. Li. *Applied Nonlinear Control*. Prentice Hall, U.S.A., 1991.

[14] Arie Levant. Robust exact differentiation via sliding mode technique. *Automatica*, 34(3):379–384, 1998.

[15] H. Huang, S. Nowoisky, R. Knoblich and C. Gühmann. Modeling and testing of the hydro-mechanical synchronization system for a double clutch transmission. 9th *Int. Modelica Conf.*, :284–294, Munich, Germany, September 2012.

[16] H. Huang, S. Nowoisky, R. Knoblich and C. Gühmann. Modeling and simulation of an automated manual transmission system. 7th *Int. Conf. on Integrated Modeling and Analysis in Applied Control and Automation* (IMAACA), :44–51, Athens, Greece, September 25–27, 2013.

[17] Silver, version 2.4.4. Product help, QTronic, 2012.

Biographies

Hua Huang works as a research assistant at the chair of electronic measurement and diagnostic technology (MDT) in the Technische Universität Berlin with the scope of model-based calibration of automated transmissions.

Clemens Gühmann (Professor) is the head of the chair of MDT in the Technische Universität Berlin. Before taking this position, he worked as an engineer at IAV automotive engineering. His research areas are modern signal processing methods (e.g. Wavelets) for automotive systems, modeling and calibration methods for automotive electronic control units as well as pattern recognition and technical diagnosis.

Yue Yu is a Master-candidate student at the department of land and sea transport systems in the Technische Universität Berlin, majoring in automotive engineering.

H. Ait-Abbas, M. Belkheiri and B. Zegnini

Adaptive Output Feedback Control For Highly Uncertain Nonlinear Systems Using Single Hidden Layer Neural Networks

Abstract: We develop an adaptive output feedback control methodology for highly uncertain nonlinear systems, in the presence of unstructured uncertainties, such as unmodelled dynamics, and unknown dimension of the regulated system. Given a smooth reference trajectory, the objective is to design a controller that forces the system measurement to track it with bounded errors. A linear in parameters neural network is introduced as an adaptive signal. A simple linear observer is proposed to generate an error signal for the adaptive laws. The network weight adaptation rule is derived from Lyapunov stability analysis, and guarantees that the adapted weight errors and the tracking error are bounded. The theoretical results are illustrated in the design of a controller for a fourth–order nonlinear system of relative degree two, and a tunnel diode circuit example having full relative degree.

Keywords: Adaptive output feedback control, nonlinear systems, unstructured uncertainties, unmodelled dynamics, linear observer, single hidden layer neural networks.

1 Introduction

A fundamental goal for many research efforts in control theory is the development of an adaptive output feedback control of uncertain nonlinear systems, mainly considering the growing interest in the use of unconventional control devices Systems that employ such devices, such as flexible robot arms, aeroelastic structures, and combustion processes, to name a few, have usually unstructured uncertainties.

Output feedback control of full relative degree systems, using a high-gain observer, was introduced in [5]. A solution to the output feedback stabilization problem for systems in which nonlinearities depend only upon the available measurement was given in [1]. An extension of these methods due to Jiang can be found in [2].

For adaptive observer design, the condition of linear dependence upon unknown parameters has been relaxed by introducing a linearly parameterized neural network (NN) in the observer structure [3]. Adaptive output feedback control using a high gain

H. Ait-Abbas, M. Belkheiri and B. Zegnini: H. Ait-Abbas[1], email: aitabbashamou@gmail.com
M. Belkheiri[2], email: zegbakeur@gmail.com B. Zegnini[1], email: mbelkhiri@yahoo.com [1] Laboratoire d'Etudes et de Développements des Matériaux Semi-conducteurs et Diélectriques, Université Amar Telidji, Algérie. [2] Laboratoire de Télécommunications, Signaux et Systèmes, Université Amar Telidji, Algérie.

De Gruyter Oldenbourg, ASSD – Advances in Systems, Signals and Devices, Volume 5, 2018, pp. 203–218.
DOI 10.1515/9783110470468-012

observer and NNs has also been proposed in [8] for nonlinear systems having input output models.

In this paper, we consider two single-input-single-output (SISO) non affine in control uncertain systems, and we propose an adaptive output feedback control methodology for both nonlinear systems using only one single-hidden-layer (SHL) NNs in order to eliminate the uncertainties. The approach employs feedback linearization, coupled with an on-line NN to compensate for modeling errors. A signal, comprised of a linear combination of the measured tracking error and the compensator states, is used to adapt the NN weights. The input vector to the NN is composed of current and past input/output data. Then, we develop a stability analysis that allows for extension of the same idea to minimum phase systems of arbitrary but otherwise bounded dimension. Numerical simulations of nonlinear systems, Van der Pol example and tunnel diode circuit model, having fourth-order nonlinear system of relative degree two and full relative degree, respectively, are used to illustrate the practical potential of the proposed approach.

2 Problem Statement

Consider the following observable nonlinear SISO system:

$$\dot{x} = f(x, u),$$
$$y = h(x). \tag{1}$$

Where $x \in \mathfrak{R}^n$ is the state of the plant, $u \in \mathbb{R}$, and $y \in \mathbb{R}$ are the input (control) and output (measurement), respectively.

*Assumption*1. The functions $f : \mathbb{R}^{n+1} \longrightarrow \mathbb{R}^n$ and $h : \mathbb{R}^n \longrightarrow \mathbb{R}$ are sufficiently smooth partially known, and the output has full relative degree r for all $(x, u) \in \Omega \times \mathbb{R}$ where $\Omega \subset \mathbb{R}^n$.

Then following [14], there exists a mapping that transforms the system in (1) into the so-called normal form:

$$\dot{\xi}_i = \xi_{i+1}, \qquad i = 1, \dots, r-1$$
$$\dot{\xi}_r = h(\xi, u)$$
$$\xi_1 = y. \tag{2}$$

where $h(\xi, u) = L_f^{(r)} h$ are the Lie derivatives, and $\xi = [\xi_1 \quad \dots \quad \xi_r]^T$. The objective is to synthesize a feedback control law that utilizes the available measurement y so that $y(t)$ tracks a smooth bounded reference trajectory $y^*(t)$ with bounded error.

3 Controller Design

3.1 Feedback Linearization

Feedback linearization is approximated by defining the following control input signal:

$$u = \widehat{h}^{-1}(y, v) \tag{3}$$

where v is referred to as a pseudocontrol. The function $\widehat{h}(y, u)$ represents the best available approximation of $h(y, u)$. Then, the system dynamics can be expressed as

$$y^{(r)} = v + \Delta \tag{4}$$

where:

$$\Delta(\xi, v) = h(\xi_1, \widehat{h}^{-1}(\xi_1, v)) - \widehat{h}(\xi_1, \widehat{h}^{-1}(\xi_1, v)) \tag{5}$$

is the inversion error. The pseudo-control is chosen to have the form

$$v = y^{*(r)} + u_d^c - u_d^a \tag{6}$$

where $y^{*(r)}$ is the rth derivative of the input signal y^*, generated by a stable command filter, u_d^c is the output of a linear dynamic compensator, u_d^a is the adaptive control signal designed to cancel Δ.

With (6), the dynamics in (4) reduce to

$$y^{(r)} = y^{*(r)} + u_d^c - u_d^a + \Delta \tag{7}$$

From (5), notice that Δ depends on u_d^a through v, whereas u_d^a has to be designed to approximately cancel Δ. Therefore, the following assumption is introduced to guarantee the existence and uniqueness of a solution for u_d^a.

Assumption 2. The mapping $u_d^a \longrightarrow \Delta$ is a contraction over the entire input domain of interest.

A contraction is defined by the condition: $|\partial\Delta/\partial u_d^a| < 1$. Using (5), this reduces to

$$\left| \frac{\partial\Delta}{\partial u_d^a} \right| = \left| \frac{\partial(h - \widehat{h})}{\partial u} \times \frac{\partial u}{\partial v} \times \frac{\partial v}{\partial u_d^a} \right| = \left| \frac{\dfrac{\partial h}{\partial u}}{\dfrac{\partial \widehat{h}}{\partial u}} - 1 \right| < 1 \tag{8}$$

Condition (8) is equivalent to the following requirements on $\hat{h}$:

$$1)\ \ \text{sign}\left(\frac{\partial \hat{h}}{\partial u}\right) = \text{sign}\left(\frac{\partial h}{\partial u}\right)$$

$$2)\ \ \infty > \left|\frac{\partial \hat{h}}{\partial u}\right| > \frac{1}{2}\left(\frac{\partial h}{\partial u}\right) > 0.$$

where the first condition means that control reversal is not allowed, and the second condition places a lower bound on the estimate of the control effectiveness in (3).

3.2 Design of the Dynamic Compensator and Tracking Error Signal Analysis

Define the output tracking error as $(\tilde{y} = y^* - y)$. Then the dynamics in (7) can be rewritten as:

$$\tilde{y}^{(r)} = -u_d^c + u_d^a - \Delta. \tag{9}$$

For the case ($\Delta = 0$), the adaptive term u_d^a is not required and the error dynamics in (9) reduce to

$$\tilde{y}^{(r)} = -u_d^c. \tag{10}$$

The following linear compensator is introduced to stabilize the dynamics in (10):

$$\begin{cases} \dot{\eta} = A_c\eta + b_c\tilde{y}, \\ u_d^c = c_c\eta + d_c\tilde{y}. \quad \eta \in \mathbb{R}^{r-1} \end{cases} \tag{11}$$

Note that η needs to be at least of dimension $(r-1)$ [4]. This follows from the fact that (10) corresponds to error dynamics that have r poles at the origin. One could elect to design a compensator of dimension $\geq r$ as well. In the future, we will assume that the minimum dimension is chosen.

Returning to (9), notice that the vector $e = [\tilde{y} \quad \dot{\tilde{y}} \quad ... \quad \tilde{y}^{(r-1)}]^T$ mutually with the compensator state η will obey the following dynamics, referred to as tracking error dynamics:

$$\begin{cases} \dot{E} = \overline{A}\,E + \overline{b}\,[u_d^a - \Delta] \\ \qquad\qquad z = \overline{C}\,E \end{cases} \tag{12}$$

where z is the vector of available measurements.

Reminder that:

$$\overline{A} = \begin{bmatrix} A - d_c bc & -bc_c \\ b_c c & A_c \end{bmatrix}, \overline{b} = \begin{bmatrix} b \\ 0 \end{bmatrix}, \overline{C} = \begin{bmatrix} c & 0 \\ 0 & I \end{bmatrix}. \tag{13}$$

and a new vector

$$E_d = \begin{bmatrix} e^T & \eta^T \end{bmatrix}^T. \tag{14}$$

where

$$A = \begin{pmatrix} 0 & 1 & 0 & \cdots & 0 \\ 0 & 0 & 1 & \cdots & 0 \\ \vdots & \ddots & \ddots & \ddots & \vdots \\ 0 & 0 & \ddots & \cdots & 1 \\ 0 & 0 & 0 & \cdots & 0 \end{pmatrix}, \quad b = \begin{bmatrix} 0 \\ 0 \\ \vdots \\ 1 \end{bmatrix}, \quad c = \begin{bmatrix} 1 \\ 0 \\ \vdots \\ 0 \end{bmatrix}^T.$$

Noting that A_c, b_c, c_c and d_c in (11) should be designed such that $\overline{A}$ is Hurwitz.

3.3 Design and Analysis of an Observer for the Error Dynamics

For the full-state feedback application [6, 7, 9, 11], Lyapunov-like stability analysis of the error dynamics in (12) results in update laws for the adaptive control parameters in terms of the error vector E. In [13, 17], and [15], adaptive state observers are used to provide the necessary estimates in the adaptation laws. However, the stability analysis was limited to second-order systems with position measurements. To rest these assumptions, we propose a simple linear observer for the tracking error dynamics in (12) and show through Lyapunov's direct method that the adaptive part of the control signal (u_d^a) can compensate for the inversion error Δ, if the output of this observer is used as an error signal for the adaptive laws.

A minimal-order observer of dimension $(r - 1)$ may be designed for the dynamics in (12). However, to streamline the subsequent stability analysis, in what follows, we consider the case of a full-order observer of dimension $(2r - 1)$ [13], [10].

To this end, consider the following linear observer for the tracking error dynamics in (12):

$$\begin{cases} \dot{\hat{E}} = \overline{A}\hat{E} + K(z - \hat{z}), \\ \hat{z} = \overline{C}\hat{E}. \end{cases} \tag{15}$$

where K is a gain matrix, and should be chosen such that $(\overline{A} - K\overline{C})$ is asymptotically stable, and z is defined in (12).

The following remark will be useful in the sequel.

REMARK 1: Equation (15) provides estimates only for the states that are feedback linearized with the transformation and not for the states that are associated with the internal dynamics.

Let

$$\widetilde{A} = \overline{A} - K\overline{C}, \quad \widetilde{E} = \hat{E} - E. \tag{16}$$

Then, the observer error dynamics can be written

$$\dot{\widetilde{E}} = \widetilde{A}\,\widetilde{E} - \overline{b}\,[u_d^a - \varDelta]. \tag{17}$$

4 Approximation of the Inversion Error

4.1 SHL NN Approximation

Usually in the literature, it has been shown that NNs are superior in their ability to compensate for modelling errors in several significant engineering applications [17], [10]. However, we will use them in controller design to cancel the effect of uncertainties due to unmodelled dynamics and unknown system parameters.

Assume that there exists an SHL NN, that approximates the inversion errors $\varDelta$. This NN has an output given by

$$y_i = \sum_{j=1}^{N_2} \left[m_{ij}\sigma\left(\sum_{k=1}^{N_1} n_{jk}x_k + \theta_{nj} \right) + \theta_{mi} \right], \quad x \in \mathbb{R}^{N_1},$$

$$i = 1, \ldots, N_3. \tag{18}$$

where $\sigma(.)$ is an activation function, n_{jk} are the first-to-second layer interconnection weights, m_{ij} are the second to third layer interconnection weights, N_2 is associated with the number of neurons in the hidden layer, θ_{nj} and θ_{mi} and are bias terms.

Such an architecture is known to be an universal approximator of continuous nonlinearities with "squashing" activation function [12], [19]. This implies that a general function $f(x) \in \mathcal{C}, x \in \mathcal{D} \subset \mathcal{R}$ can be written as:

$$f(x) = M^T\sigma(N_0^T x) + \varepsilon(x). \tag{19}$$

where $\varepsilon(x)$ is the function reconstruction error.In general, given a constant real number $\varepsilon^* > 0, f(x)$ is within $\varepsilon^* > 0$ range of the NN (18), if there exist constant weights M and N, such that for all $x \in \mathcal{D} \in \mathcal{R}$, the representation in (19) holds with $\|\varepsilon\| < \varepsilon^*$.

The following theorem extends these results to map the unknown dynamics of an observable system from available input–output history [13].

<u>Theorem</u> 2: Given a compact set $\mathcal{D} \subset R^{n+1}$ and $\varepsilon^* > 0$, the model inversion error $\Delta(\xi, v)$ can be approximated over $\mathcal{D}$ by an SHL NN

$$\Delta(x, u) = M^T \sigma(N_0^T \mu) + \varepsilon(d, \mu), \quad |\varepsilon| < \varepsilon^*. \tag{20}$$

Using the input vector

$$\mu(t) = [1 \quad \bar{v}_d^T(t) \quad \bar{y}_d^T(t)]^T \in \mathcal{D}, \quad \|\mu\| \le \mu^*, \quad \mu^* > 0. \tag{21}$$

where the first component is introduced to approximate a nonzero offset of Δ. Then, note that

$$\bar{v}_d^T(t) = [v(t) \quad v(t-d) \quad \dots \quad v(t-(n_1-r-1)d)]^T,$$
$$\bar{y}_d^T(t) = [y(t) \quad y(t-d) \quad \dots \quad y(t-(n_1-1)d)]^T.$$

with $n_1 \ge n, d > 0$ denoting time-delay and μ^* being a uniform bound for all $(x, u) \in \mathcal{D}$.

4.2 Adaptive Control

The adaptive signal is chosen to be the output of an SHL NN

$$u_d^a = \widehat{M}^T \sigma(N_0^T \mu) + \varepsilon(d, \mu). \tag{22}$$

where $\widehat{M}$ is the estimate of M that is updated according to the following adaptation law:

$$\dot{\widehat{M}} = -F[2(\hat{\sigma} - \hat{\sigma}\prime N_0^T \mu)\widehat{E}^T P\bar{b} + k(\widehat{M} - M_0)]. \tag{23}$$

in which M_0 is the initial value of M, N_0 is the initial value of the hidden layer weights vector, $\hat{\sigma} = \sigma(N_0^T \mu_d)$, $\hat{\sigma}\prime$ denotes the Jacobian matrix, P is the solution of the Lyapunov equation

$$\bar{A}^T P + P\bar{A} = -Q. \tag{24}$$

for some $Q > 0$, $k > 0$, and F is the adaptation gain matrices.

Notice that in (22), there is an algebraic loop, since μ, by definition, depends upon u_d^a through v, see (21). However, with bounded squashing functions, this algebraic loop has at least one fixed-point solution.

Using (20) and (22), the error dynamics in (12) can be formulated as

$$\begin{cases} \dot{E} = \bar{A}\,E + \bar{b}\,[\widehat{M}^T \sigma(N_0^T \mu) - M^T \sigma(N_0^T \mu) - \varepsilon], \\ z = \bar{C}\,E. \end{cases} \tag{25}$$

Define

$$\widetilde{M} = \widehat{M} - M, \quad \tilde{z}_d = \begin{bmatrix} \widetilde{M}^T & 0 \end{bmatrix}^T. \tag{26}$$

and note that

$$\|\widehat{M}\| < \|\widetilde{M}\| + M^*. \tag{27}$$

where M^* is the upper bound for the weights in (20)

$$\|\widehat{M}\| < M^*. \tag{28}$$

Since the inner layer weights N_0 are adjusted off-line to fit the real weights N, then we can write the mismatch between the adaptive signal and the real NN as :

$$u_d^a - \Delta = \widehat{M}^T \sigma(N_0^T \mu) - M^T \sigma(N_0^T \mu) - \varepsilon. \tag{29}$$

With (27), the previous representation (29), allows for the following upper bound for some computable α_1, α_2:

$$|u_d^a - \Delta| \le \alpha_1 \|\widehat{Z}_d\|_F + \alpha_2, \quad \alpha_1 > 0, \quad \alpha_2 > 0. \tag{30}$$

The subscript F denoting the Frobenius norm.

For the stability proof, we will need the following representation

$$\widehat{M}^T \sigma(N_0^T \mu) - M^T \sigma(N_0^T \mu) = \widetilde{M}^T (\hat{\sigma} - \hat{\sigma}\prime N_0^T \mu)$$
$$+ \widetilde{M}^T \hat{\sigma}\prime N_0^T \mu + w_d.$$

where

$$w_d = \widetilde{M}^T \hat{\sigma}\prime N_0^T \mu.$$

With the bound in (21), a bound for $(w_d - \varepsilon)$ over a compact set can be presented as follows [7]:

$$|w_d - \varepsilon| \le \gamma_1 \|\widetilde{Z}_d\|_F + \gamma_2, \quad \gamma_1 > 0, \quad \gamma_2 > 0. \tag{31}$$

where γ_1 and γ_2 are computable constants, γ_1 depends upon unknown constant μ^*, and γ_2 upon ε^*. Thus, the forcing term in (25) can be rewritten as

$$u_d^a - \Delta = \widetilde{M}^T (\hat{\sigma} - \hat{\sigma}\prime N_0^T \mu) + M^T \hat{\sigma}\prime N_0^T \mu + w_d - \varepsilon. \tag{32}$$

5 Stability Analysis

In the current section, we confirm through Lyapunov's direct method that if the initial errors of the variables E^T, $\tilde{E}^T$, $\tilde{M}$ and belong to a prescribed compact set, then they are ultimately bounded.

For stability analysis, introduce the observer error signal $\tilde{E} = \hat{E} - E$. The observer error dynamics can then be written as:

$$\dot{\tilde{E}} = \tilde{A}\,\tilde{E} - \bar{b}\,[\tilde{M}^T(\hat{\sigma} - \hat{\sigma}\prime N_0^T \mu) + M^T \hat{\sigma}\prime N_0^T \mu + w_d - \varepsilon] \tag{33}$$

where

$$\tilde{A} = \bar{A} - K\bar{C}$$

<u>LEMMA</u>: Following [16], if A is any asymptotically stable matrix, then given any positive definite symmetric matrix $Q > 0$, there exists a unique positive definite symmetric matrix $P > 0$ such that

$$\bar{A}^T P + P\bar{A} = -Q.$$

<u>REMARK</u>: Consider the compact set D and initial conditions in Ω_α. The feedback control law given by

$$u = \hat{h}^{-1}(y, v)$$

where

$$v = y^{*(r)} + u_d^c - u_d^a$$
$$u_d^c = c_c \eta + d_c e$$
$$u_d^a = \widehat{M}^T \sigma(\widehat{N}_0^T \mu)$$

with the following adaptive laws

$$\dot{\widehat{M}} = -F[2(\hat{\sigma} - \hat{\sigma}\prime N_0^T \mu)\hat{E}^T P\bar{b} + k(\widehat{M} - M_0)].$$
$$\dot{\hat{\psi}} = \gamma[2\hat{E}^T P_2\bar{b}\,\text{sign}\,(2\hat{E}^T P_2\bar{b}) - \lambda_\psi(\psi - \psi_0)]$$

For $F, \gamma > 0$, $\lambda_\psi > 0$, $\psi_0 > 0$ is initial value of the estimate ψ, P_2 is the solution to the following Lyapunov equation

$$\bar{A}^T P_2 + P_2\bar{A} = -Q_2.$$

for some $Q_2 > 0$, guarantees that all the error signals in the closed loop system are ultimately bounded (refer to [18] for more details).

6 Application

To illustrate the performance of the proposed adaptive controller in the presence of unstructured uncertainties, we consider the two nonlinear systems, having a fourth-order nonlinear system of relative degree two (Van der Pol) and the tunnel diode circuit example with full relative degree.

6.1 Tunnel Diode Circuit Example

$$\begin{cases} \dot{x}_1 = \dfrac{1}{C}x_2 - \dfrac{1}{C}h(x_1) \\ \dot{x}_2 = -\dfrac{R}{L}x_2 - \dfrac{1}{L}x_1 + \dfrac{u}{L} \end{cases} \tag{34}$$

where x_1 the voltage across the capacitor C and x_2 is the current through the inductor L. The initial conditions were set as $x_1(0) = 0.1, x_2(0) = 0.0005$, and the element values of the circuit are $R = 1.5k\Omega, L = 1nH$, and $C = 2pF$.

The function $h : \mathbb{R} \longrightarrow \mathbb{R}$ represents the characteristic curve of the tunnel diode,

$$h(x_1) = x_1 + 2x_1^2 + x_1^3 - x_1^4 - 2x_1^5 \tag{35}$$

The output y has a full relative degree of $n = r = 2$.

6.2 Van der Pol Example

$$\begin{cases} \dot{x}_1 = x_2 \\ \dot{x}_2 = -0.2(x_1^2 - 1)x_2 - 0.2x_3 + \dfrac{u}{\sqrt{|u| + 0.1}} \\ \dot{x}_3 = x_4 \\ \dot{x}_4 = -0.2x_4 - x_2 + x_1 \end{cases} \tag{36}$$

with initial conditions $x_1(0) = 0.5, x_2(0) = 1.5, x_3(0) = 0$ and $x_4(0) = 0$. The output y has a relative degree of $r = 2$.

The following dynamic compensator:

$$\begin{cases} \dot{\eta} = -5\eta + 5\tilde{y} \\ u_d^c = -6.8\eta + 8\tilde{y} \end{cases} \tag{37}$$

places the poles of the closed-loop error dynamics in (10) of both nonlinear systems at $-3, -1 \pm j$, so that

$$Q_{desired}(P) = Q_{\overline{A}}(P) \qquad (38)$$

where $Q_{desired}(P)$ is the characteristic polynomial of the desired poles, whereas $Q_{\overline{A}}(P)$ is the characteristic polynomial of $\overline{A}$.

The observer dynamics in (33) were designed so that its poles are four times faster than those of the error dynamics. We implemented seven neurons in the hidden layer, and the following sigmoidal basis function:

$$\sigma(x) = \frac{1}{1 + e^{-ax}} \qquad (39)$$

with $a = 1$. The adaptation gains were set to $F = 2I$, with sigma modification gain $k = 0.65$.

The contribution of this paper is to design only one single hidden layer neural network (SHL NN) that compensate adaptively for the nonlinearities of both systems Van der Pol example and tunnel diode circuit model, what bring to force the system measurement to track reference trajectory with bounded errors. First, setting the output $y = x_1$ for each system. However, the reader is reminded that the controller has been designed given only the fact that $r = 2$. Taking advantage of this fact, we employ feedback linearization, coupled with an on-line NN to compensate for modeling errors, according to the equation (7). The fixed structure of a dynamic compensator, described in (11) and (37), is designed to stabilize the linearized system. A signal, comprised of a linear combination of the measured tracking error and the compensator states, presented in (25), is used to adapt the NN weights [10].

Figure 1 compares the system response y without NN augmentation (dashed line) with the reference model output y^* (solid line), clearly demonstrating the almost unstable oscillatory behavior caused by the nonlinear elements (Δ) in the Van der Pol model in the first half time (0 to 40 seconds) and the nonlinearities of the tunnel diode equation in the last half time (40 to 80 seconds). While, with the aid of NN augmentation, Fig. 2 shows that these oscillations are eliminated after a period of about $1s$. This is accounted for the successful identification of the model inversion error (Δ) (solid line) by the adaptive output signal (u_d^a) (dashed line), which is illustrated in Fig. 3.

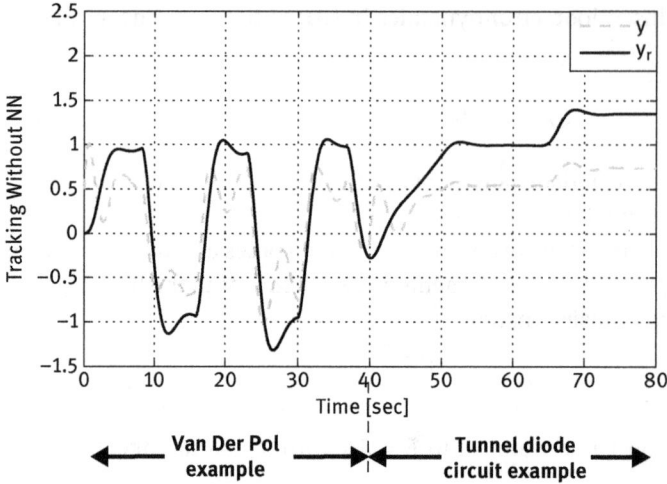

Fig. 1. Tracking without NN.

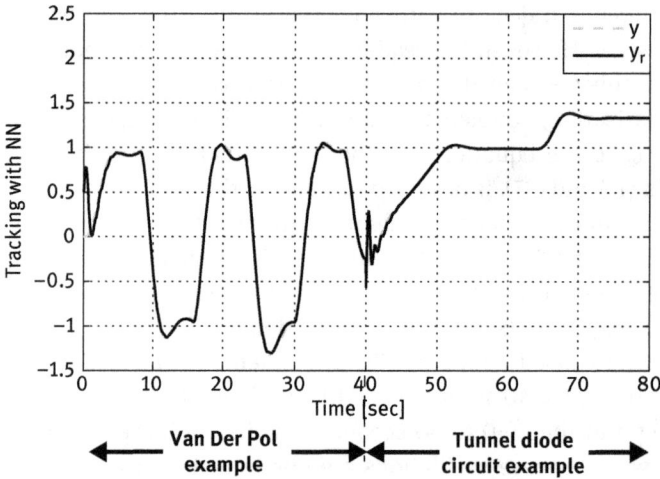

Fig. 2. Tracking with NN.

Figure 4 compares the control efforts $(y^* - y)$ without and with adaptation, where the NN based adaptive controller (u_d^a) exhibits a steady state tracking error. This error can be minimized when designing an excellent linear compensator, according to the equation (9).

The NN controller weights history are shown in Fig. 5. Moreover, one chooses a good structure of the network of neuron in order to avoid the phenomenon of on training which deteriorates the architecture.

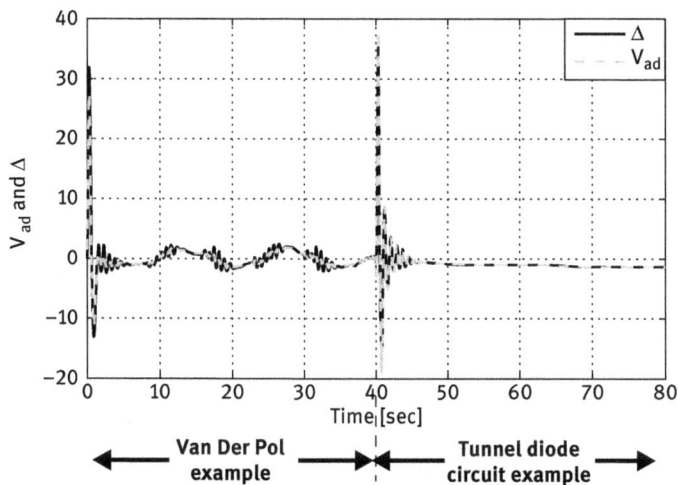

Fig. 3. Identification of uncertainties (Δ) by NN (u_d^a).

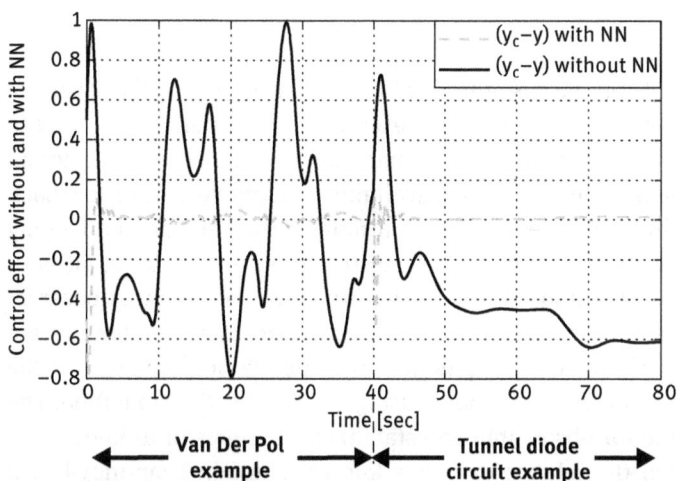

Fig. 4. Control effort without and with NN.

As expected, the SHL NN improves the tracking performance due to its ability to "model" nonlinearities, even with the unmodelled dynamics. Consequently, simulations show that the NNs augmented adaptive output feedback controller compensates successfully for high unstructured uncertainties.

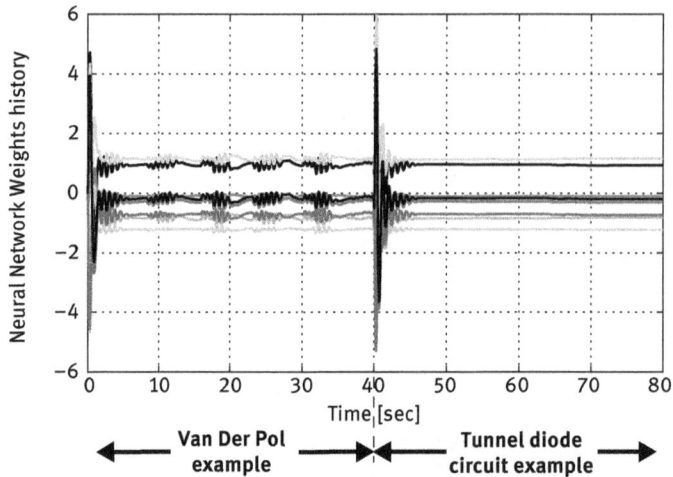

Fig. 5. NN weights history.

6.3 Summary

In this paper, a new control approach is proposed for adaptive output feedback control of uncertain nonlinear systems using NNs. However, the control performance of the given nonlinear systems still influenced by the unmodelled dynamics and/or external disturbances, and to compensate for these uncertainties adaptive output feedback control is proposed. In this note, the obtained controller is then augmented by only one SHL NN used as an approximator for the unstructured uncertainties in both nonlinear systems Van der Pol and Tunnel Diode Circuit. A simple linear observer is introduced to estimate the derivatives of the tracking error. These estimates are used in the adaptation laws for the NN parameters. Ultimate boundedness of the tracking error and observation error are shown using Lyapunov's direct method. The methodology is applicable for observable and stabilizable systems of unknown but bounded dimension when the relative degree is known. Through Lyapunov-based theoretical analysis, computer simulation, we were able to demonstrate that the proposed neural network-based adaptive output feedback controller was robust to modeling inaccuracies, and excellent tracking performance was succeeded.

Bibliography

[1]   A. Praly and Z. P. Jiang. Stabilization by output feedback for systems with ISS inverse dynamics. *Syst. Control Lett.*, 21:19–33, 1993.
[2]   Z. P. Jiang. A combined backstepping and small-gain approach to adaptive output feedback control. *Automatica*, 35:1131–1139, 1999.

[3]   F. L. Lewis, K. Liu, and A. Yesildirek. Control of Robot Manipulators. *New York, Macmillan*, 1993.
[4]   J. Brasch and J. Pearson. Pole placement using dynamic compensators. *IEEE Trans. Automat. Contr.*, AC-15:34–43, Jan. 1970.
[5]   F. Esfandiari and H. K. Khalil. Output feedback stabilization of fully linearizable systems. *Int. J. Control*, 56(5):1007–1037, 1992.
[6]   A. J. Calise, S. Lee, and M. Sharma. Development of a reconfigurable flight control law for a tailless aircraft. *J. Guidance, Contr., Dynamics*, 24(5):896–902, 2001.
[7]   F. L. Lewis, A. Yesildirek, and K. Liu. Multilayer neural-net robot controller with guaranteed tracking performance. *IEEE Trans. Neural Networks*, 7:1–12, Jan. 1996.
[8]   S. Seshagiri and H. K. Khalil. Output feedback control of nonlinear systems using RBF neural networks. *IEEE Trans. Neural Networks*, 11:69–79, Feb. 2000.
[9]   M. B. McFarland and A. J. Calise. Multilayer neural networks and adaptive control of agile anti-air missile. *Proc. AIAA Guidance, Navigation, Contr. Conf., AIAA Paper 97–3540*, 1997.
[10] H. Ait Abbas, M. Belkhiri and B. Zegnini. Feedback Linearization Control of a Class of Nonlinear Uncertain Systems Using Neural Networks. *International Journal on Advanced Electrical Engineering*, 01(02):121–131, 2013.
[11] F. Nardi and A. J. Calise. Robust adaptive nonlinear control using single hidden layer neural networks. *Proc. Conf. Decision Contr.*, pp. 3825–3830, 2000.
[12] K. Funahashi. On the approximate realization of continuous mappings by neural networks. *Neural Networks.*, 2:183–192, 1989.
[13] N. Hovakimyan, F. Nardi, A. J. Calise, and H. Lee. Adaptive output feedback control of a class of nonlinear systems. *Int. J. Contr.*, 74(12):1161–1169, 2001.
[14] A. Isidori. Nonlinear Control Systems. *Berlin, Germany: Springer-Verlag*, 1995.
[15] Y. Kim and F. L. Lewis. High Level Feedback Control with Neural Networks. *Singapore: World Scientific*, 1998.
[16] A. J. Calise, N. Hovakimyan, and M. Idan. Adaptive output feedback control of nonlinear systems using neural networks. *Automatica.*, 37(8):1201–1211, 2001.
[17] N. Hovakimyan, R. T. Rysdyk, and A. J. Calise. Dynamic neural networks for output feedback control. *Proc. Conf. Decision Contr.*, pp. 1685–1690, 1999.
[18] Flavio Nardi. Neural Network based Adaptive Algorithms for Nonlinear Control. *A Thesis Presented to The Academic Faculty of The School of Aerospace Engineering.*, Georgia Institute of Technology, November 2000.
[19] K. Hornik, M. Stinchcombe, and H. White. Multilayer feedforward networks are universal approximators. Neural Networks., 2:359–366, 1989.

Biographies

Hamou Ait-Abbas received his License and Master degrees in Electrical Machines from University M'hammed Bouguerra (UMBB), Boumerdes, Algeria in 2009 and 2011, respectively. He is currently working towards a Ph.D. degree in control of electromechanical systems at the University Amar Telidji, Laghouat, Algeria. He is working as an assistant teacher in electrical engineering department at University of Amar Telidji, Laghouat, Algeria. His research interests include theory and applications of nonlinear control and, more recently, nonlinear adaptive control.

Mohammed Belkheiri was born on November 1977 in Djelfa, Algeria. He received the Engineer degree in electrical engineering from University of Boumerdes, Algeria, in 2000, the Master degree in robotics and automatic control from Military Polytechnic School, Bordj El-bahri, Algiers, in 2002, and the Doctorate degree in Automatic Control from National Polytechnic School of Algiers, ENP Algeria since 2003. He is working as an assistant professor at the department of electrical engineering at University of Amar Thelidji, Laghouat, Algeria. His research interests include nonlinear and adaptive control and applying neural networks for solving complex engineering problems.

Boubakeur Zegnini was born on 25/01/1968. He received the applied electrical engineering degree from ENSET Laghouat, Algeria ,in1991, the M.Sc. degree from University of Laghouat in 2001. He received his Ph.D. degree in Electrical Engineering from University of science and Technology USTO, Oran, Algeria in 2007. He was professor of technical secondary school from 1991 to 2001. Since 2001, he is working as an associate professor with the Department of Electrical Engineering at University of Laghouat, Algeria. He joined the Laboratory of Electrical Engineering at Paul Sabatier University of Toulouse, France, "solid dielectrics and reliability" research team from 2005 to 2007. Currently, he is head of research team in the Dielectric materials Laboratory LeDMaScD at the University Amar Telidji of Laghouat, in Algeria. Following this, he became a Full Professor at University Amar Telidji of Laghouat, in Algeria.His main research interests include high voltage, dielectric materials, outdoor insulation, numerical modeling and simulation. He is author and co-authors of many scientific publications.

M. Taktak-Meziou, A. Chemori, J. Ghommam and N. Derbel

RISE Feedback with NN Feedforward Control of a Servo-Positioning System for Track Following in HDD

Abstract: This paper addresses design challenges associated with a servo system of a Hard Disc Drive (HDD). The recently developed Robust Integral of Sign Error (RISE) approach is proposed to control the Read/Write (R/W) head tip of the HDD. Such a technique, combined with a feedforward Neural Network (NN) control term, is not only able to meet the different imposed constraints on the system, but also guaranty the asymptotic stability of the overall closed-loop system. To the best knowledge of the authors, the proposed controller, applied at the low frequency region of a HDD, has never been conducted before on such a system. A comparative study between the RISE-NN controller and the classical Proportional Integral Derivative (PID) is performed under various operating conditions ranging from nominal case without external disturbances to more complex cases with disturbances and parametric uncertainties. The main objective of this study is to highlight the effectiveness of RISE-NN control approach in solving the track following problem in HDD.

Keywords: RISE feedback control, nonlinear systems, hard-disc-drives, asymptotic stability, robust control, neural networks.

1 Introduction

Since its appearance in 1956, the Hard-Disc-Drive (HDD) technology has continued to progress over the years. This accelerated evolution has primarily affected the storage capacity growing from few Megabytes to several Terabytes. In addition, many other characteristics have undergone perceptible changes [1]. These include the size factor which has been reduced from 24-inch diameter in a 50 disks prototype to one disk of 3.5-inch diameter, the spindle speed which has steadily increased to reach 15000 rpm in the latest versions of HDDs,etc. The main objective of such a development is to improve the system operating performance in terms of access speed, precision of the positioning, data access, and reliability.

M. Taktak-Meziou, A. Chemori, J. Ghommam and N. Derbel: [1] Research unit on Control & Energy Management Laboratory (CEMLab), University of Sfax, National School of Engineers of Sfax, Sfax, Tunisia, email: manel.taktak@yahoo.com, email: jawhar.ghommam@gmail.com, email: n.derbel@enis.rnu.tn. [2] Laboratoire d'Informatique, de Robotique et de Microélectronique de Montpellier (LIRMM), France, email: chemori@lirmm.fr.

De Gruyter Oldenbourg, ASSD – Advances in Systems, Signals and Devices, Volume 5, 2018, pp. 219–240.
DOI 10.1515/9783110470468-013

From a technical standpoint, it can clearly be seen that a HDD is a mechatronic system. Indeed, the system is constructed through the synergistic integration of mechanical engineering, electronics, control engineering, and computer technology. Figure 1 shows a view of the main components of a typical HDD servo-system.

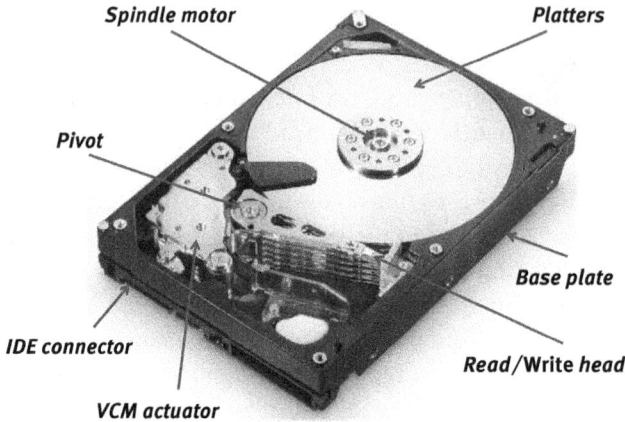

Fig. 1. View of the main components of a typical HDD.

A HDD is composed of a spindle motor devoted to drive the rotating platters, where digital data are stored in concentric tracks. In order to treat these data, whether to read from or write on the disc, the system is equipped with several magnetic Read/Write (R/W) heads. These heads are connected to a second motor, called Voice-Coil-Motor (VCM), which is designed to manage their movement on the disk surface and achieve access to the desired track.

A good HDD is evaluated according to its ability to move the R/W head tip rapidly from its current position to a desired target track and to maintain it as close as possible to its center while treating data. Therefore, the Position Error Signal (PES), defined as the deviation of the R/W head from the desired track center, should be as minimal as possible in order to guarantee a reliable data reading or writing. Such a regulation is tighter with the modern HDD servo-systems becoming more and more small in size. In addition to the reduced size factor, many sources of errors can be noticed in the system. These factors contribute significantly to the degradation of the overall system's performance in terms of precision and access to the information. They mainly include nonlinear frictions caused by the pivot bearing and the flex cable and inaccuracies caused by the movement of the head form one track to another. Generally, a HDD is often subject to various disturbances which can be classified into three categories: The input disturbances caused by mechanical perturbations such as resonances, friction and vibrations The output disturbances which are due to the rotation of the spindle motor rotation and its effects. The measurement noise caused by the position-measurement techniques and/or sensors.

All the above-listed errors' sources threaten the performances of the HDD servo-system and may degrade the system reliability. Consequently, it would be necessary to deal with them rigorously and compensate their effects as much as possible.

To do that, several research efforts have been devoted to design efficient robust controllers. Their common objective was not only to overcome the different HDD problems cited above, but also to ensure a tighter PES while positioning the R/W head even in the presence of eventual disturbances, nonlinearities, and inaccuracies on the system's dynamics.

Among these control solutions, we can distinguish classical approaches such as PID controllers [2], lead-lag compensator [3] and classical filters [4] which can no longer meet the demand for HDDs higher performances. Accordingly, to deal with these difficulties, several control attempts have been recently developed including (i) advanced control approaches such as optimal controllers [5, 6], Composite Nonlinear Feedback technique (CNF) [7] and Robust Perfect Tracking (RPT) [8] and (ii) several robust control solutions such as adaptive control [9, 10], sliding mode control [11, 12], robust control [13, 14] and lately predictive control approaches [15, 16]. Some of these methods have been experimentally tested to show their strengths and weaknesses on a real system.

This paper is dedicated to the application of the recently developed control method based on Robust Integral Sign of the Error (RISE) [17] to the case of HDDs. This control technique is chosen based on its advantages in addressing the problem of trajectory tracking of a class of uncertain and high order nonlinear systems [17, 18]. Since RISE is a high gain feedback method, the idea proposed in [19] was to develop an improved version of this technique which involves the combination of a Neural Network based feedforward control term with the feedback controller. In this paper, this control solution is proposed to address the track following problem in a HDD servo-system and to ensure both robust performance and asymptotic stability of the overall closed-loop servo-system.

The reminder of this paper is organized as follows. In section 2, the HDD low-frequencies dynamic modeling is introduced. Then, in section 3, the HDD servo-positioning control problem is formulated. In section 4, the RISE feedback based NN controller is developed. Section 5 is devoted to a comparative study between the proposed RISE-NN and a classical PID controllers, where numerical simulations in different operating conditions are presented and discussed. Finally, in section 6, some concluding remarks are drawn.

2 HDD Low Frequencies Dynamic Modeling

In a HDD servo-system, one of the important limitations for high track density is the nonlinear effects arising from frictions. Such nonlinear frictions are mainly induced by the pivot bearing and data flex cable in the VCM actuator (see Fig. 1). Their presence leads to the generation of large residual errors and oscillations which degrade the overall system performances and reduce its reliability.

Certainly, a deeply understanding of the nonlinear friction behavior would be helpful to find an efficient control solution that mitigate their degrading effects. Therefore, for the aim of developing a representative friction model in a HDD servo-system, many researchers have provide considerable efforts in the literature [20, 21]. The best representation that encompasses all static and dynamic features turned out to be that of LuGre friction model [22]. For a complete review of the friction modeling, the reader is referred to [23].

In order to enhance the track following performances in a HDD servo-positioning system, it would be necessary to compensate the overall nonlinear frictions. A survey of the literature showed that many control approaches dealing with the above compensation have been proposed. Some of them include an accurate modeling of the friction behavior [24, 25] and others are non-model-based friction estimation [26, 27]. In spite of all the proposed solutions, the study of the nonlinear friction behavior and the search for a good compensation control solution are still open problems in HDD technology.

Based on recent works of [28], the low-frequency mathematical model of the VCM actuator can be expressed as follows:

$$M(q)\ddot{q} + F(q, \dot{q}) = u \tag{1}$$
$$y = q + w_{out}$$

where $M(q)$ denotes the system inertia verifying $M(q) > 0$. q, $\dot{q}$ and $\ddot{q}$ denote the actual position, velocity and acceleration of the VCM-actuator head tip respectively. u represents the control input, y is the measured position of the VCM-actuator in presence of the eventual output disturbance w_{out} representing external vibrations and chocks caused by the flexibility of the material. $F(q, \dot{q})$ is a nonlinear function representing the pivot bearing hysteresis friction whose behavior can be described by the LuGre friction model [22]. This last one is expressed as follows:

$$F(q, \dot{q}) = \sigma_0 z + \sigma_1 \dot{z} + \sigma_2 \dot{q} \tag{2}$$
$$\dot{z} = \dot{q} - \alpha(\dot{q}) \, | \, \dot{q} \, | \, z \tag{3}$$
$$\alpha(\dot{q}) = \frac{\sigma_0}{f_c + (f_s - f_c) \, \exp\left(-\left[\dfrac{\dot{q}}{\dot{q}_s}\right]^2\right)} \tag{4}$$

where z is an internal state of the friction model assumed to be immeasurable. σ_0, σ_1, and σ_2 are the model parameters reflecting the small displacements which are the stiffness, the micro damping, and viscous coefficient respectively. f_s corresponds to the stiction force, f_c is the Coulomb friction force, and the parameter q_s is the Stribeck velocity [29].

3 Control Problem Formulation

The main goal of a HDD servo-system control is to read/write data from/on concentric track circumscribed onto the disc surface. Therefore, by controlling the current in the VCM, the head is able to move in both directions to follow the desired target track. Consequently, in order to reach its target, two main functioning modes can be distinguished [1]: The first one, being the *track seeking* mode, deals with moving the R/W head from one desired track to another, at a distance of about a micro-inch between two adjacent tracks. The displacement of the head is required to be as quick as possible with a limited control effort. The second mode, being the *track following* mode, consists of maintaining the head as close as possible to the center of the desired data track to guarantee an accurate positioning, crucial for reading/writing digital data. Therefore, the drive initiates its functioning by a track seeking control with a saturated control law. Then, when the head is positioned onto the target track, the drive switches into the track following mode. A schematic illustration of the above mode functions is shown in Fig. 2.

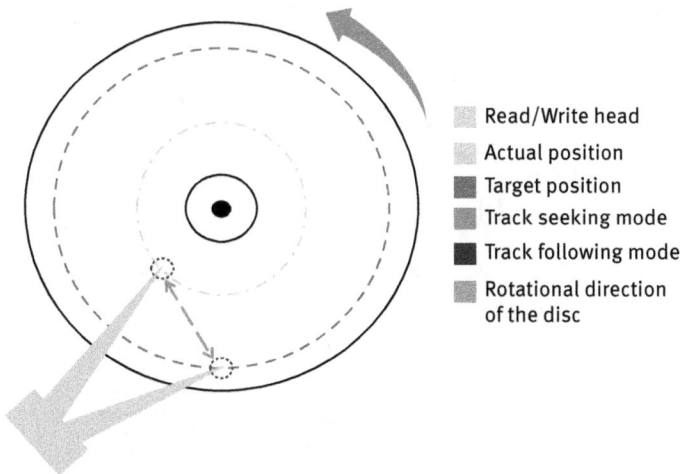

Fig. 2. Illustration of the main operating functions of a HDD servo-system.

Let q_d be the desired track position. In order to evaluate the tracking performance, the position error e_1 is introduced. It is defined as the deviation of the HDD head tip from the center of the desired position:

$$e_1 = q_d - q \tag{5}$$

The main control objective consists of moving the head onto the surface of the disc so that it follows a predefined target track. Then, the head is required to be as close as possible to this desired position while reading/writing data, in order to ensure superior HDD performances. The setting equation of the control objective can be reformulated as follows:

$$\lim_{t \to \infty} |e_1(t)| = \lim_{t \to \infty} |q_d(t) - q(t)| = 0 \tag{6}$$

In the present paper we aim to design an efficient control solution for the track-following mode.

4 Proposed Control Solution

The recent developed feedback control strategy RISE [17] is proposed in this paper to deal with the track following problem of the HDD servo-system. Such a control technique, blend with a NN-based feedforward, is able to deal with the non-explicit knowledge of the friction model $F(q, \dot{q})$ introduced in the dynamic model (1)-(4). Before going further, to give an overview of the control strategy, an illustrative block diagram of such controller is shown in Fig. 3.

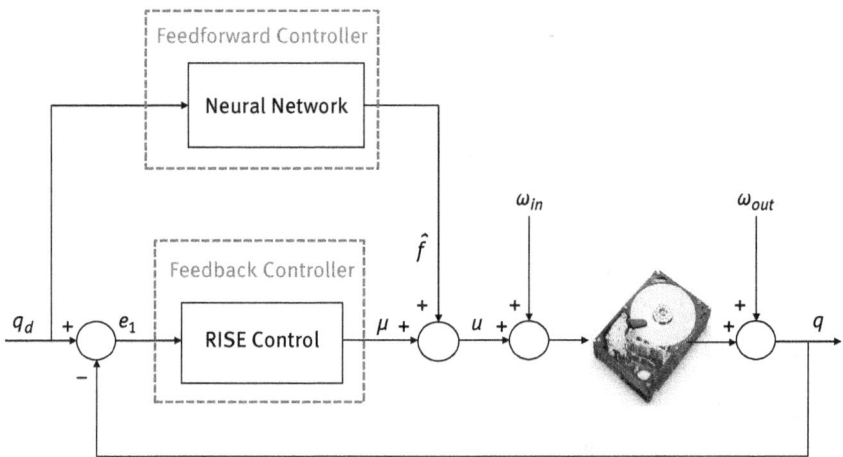

Fig. 3. Overview of the control scheme including a RISE feedback with a NN feedforward.

Starting from a desired target track q_d, the global control input u, which is the sum of a feedforward NN control term $\hat{f}$ and a RISE feedback term μ, is calculated at each sample time to move the head tip to the desired position. In the following section, a background on NN-based Feedforward then RISE Feedback controllers are introduced illustrating how they can be combined together to achieve an asymptotic stability of the overall closed-loop system.

4.1 Background on NN Feedforward Control

Dynamic neural networks present an effective tool for estimation and control of nonlinear and complex systems [30]. The universal approximation remains the key feature of the NN-based controllers [31]. Consider $\mathbb{S}$, a compact set and $f(x)$ a smooth function defined as $f : \mathbb{S} \rightarrow \mathbb{R}^n$. There exists always three-layer NN able to represent $f(x)$ [19] such that $f(x) = W^{\mathsf{T}} \sigma(V^{\mathsf{T}} x) + \varepsilon(x)$ for given inputs $x(t) \in \mathbb{R}^{a+1}$. $V \in \mathbb{R}^{(a+1) \times L}$ is a bounded constant weight matrix for the first-to-second layer and $W \in \mathbb{R}^{(L+1) \times 1}$ is the ideal weight matrix for the second-to-third layer. a is the number of inputs and L is the number of neurons in the hidden layer. $\sigma(.) \in \mathbb{R}^{L+1}$ is the activation function and $\varepsilon(x) \in \mathbb{R}^n$ is the functional error approximation, satisfying $\| \varepsilon(x) \| \leq \varepsilon_N$, with ε_N is a known constant bound. Figure 4 shows an illustrative description of a three-layer NN principle.

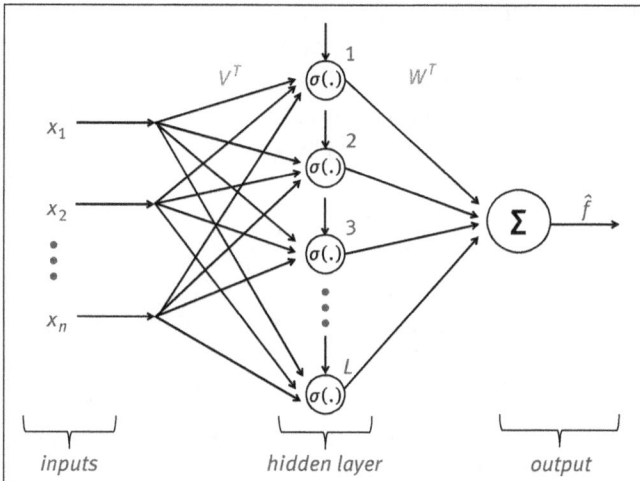

Fig. 4. Schematic view of a three-layer NN.

<u>REMARK 1</u>: The activation function $\sigma(.)$ can take different forms such as sigmoid, hyperbolic tangent or a radial basis function. In this paper, the considered $\sigma(.)$ is a radial basis function described by the following equation:

$$\sigma(x_i) = \exp\left(-\frac{\|x_i - c_i\|^2}{\sigma_i^2}\right) \ , \quad \forall i \in \mathbb{N}$$

where c_i is the center of the basis function and σ_i is its width; they are chosen a priori and kept fixed throughout this work for reason of simplicity.

For subsequent developed calculations of the control input, some assumptions and properties have to be exploited which are the following:

ASSUMPTION The desired position q_d, as well as its first and second time derivatives exist and are all bounded, i. e., q_d, $\dot{q}_d$, and $\ddot{q}_d \in \mathcal{L}_\infty$.

PROPERTY The NN quantities are bounded such as $\| W \| \le W_m$, $\| \sigma \| \le \sigma_m$, where W_m and σ_m are known positive constants [32].

4.2 Background on RISE Feedback Control

In this work, the main control objective is to maintain the R/W head as close as possible to a predefined desired position in order to perform an accurate track following task. Therefore, a RISE feedback control approach with NN feedforward estimation is therefore proposed as a control solution (i) to deal with the unknown nonlinear dynamics and (ii) to guarantee the asymptotic stability of the controlled HDD model (1)-(4). The control strategy is detailed in this section, introducing the open-loop and closed-loop tracking errors. Based on assumption 1, the position tracking error $e_1(t)$, the filtered tracking errors denoted by $e_2(t)$ and $r(t)$, are defined as follow

$$e_1 = q_d - q \tag{7}$$

$$e_2 = \dot{e}_1 + \alpha_1 e_1 \tag{8}$$

$$r = \dot{e}_2 + \alpha_2 e_2 \tag{9}$$

where α_1 and α_2 are positive tuning gains.

<u>REMARK 2</u>: The filtered tracking error $r(t)$ is an immeasurable quantity since it depends on $\ddot{q}(t)$ which is not measurable in a HDD.

4.2.1 Open-loop tracking error system

To develop the open-loop tracking error system, a multiplication of (9) by $M(q)$ is made. Then, based on the expressions (1), (7), and (8), the resulting system can be expressed as follow:

$$M(q)r = F_d + S - u \tag{10}$$

where F_d is an auxiliary function defined by:

$$F_d = M(q)\ddot{q}_d + F(q_d, \dot{q}_d) \tag{11}$$

and S is a second auxiliary function defined by:

$$S = M(q)(\alpha_1 \dot{e}_1 + \alpha_2 \dot{e}_2) + F(q, \dot{q}) - F(q_d, \dot{q}_d) \tag{12}$$

Based on the NN approximation, F_d can be expressed as follows:

$$\dot{F}_d = W^{\top} \sigma(V^{\top} x_d) + \varepsilon(x_d) \tag{13}$$

where $x_d = \begin{bmatrix} 1 & q_d & \dot{q}_d & \ddot{q}_d \end{bmatrix}^{\top}$ and $\varepsilon(x_d)$ is the bounded NN approximation error. According to assumption 1, the following inequalities hold:

$$\| \varepsilon(x_d) \| \leq \varepsilon_N \tag{14}$$

$$\| \dot{\varepsilon}(x_d, \dot{x}_d) \| \leq \varepsilon'_N \tag{15}$$

where ε_N and ε'_N are known positive bounded constants.

4.2.2 Closed-loop tracking error system

Using the previous open-loop tracking error system (10), the control input can be expressed as the sum of the feedforward NN estimation term and the RISE feedback term. As detailed in [33], the RISE control term $\mu(t)$ is given by:

$$\mu(t) = (k_s + 1)e_2(t) - (k_s + 1)e_2(0) + \int_0^t \{(k_s + 1)\alpha_2 e_2(s) + \beta_1 \operatorname{sign}[e_2(s)]\} ds \tag{16}$$

where $k_s, \beta_1 \in \mathbb{R}^+$ are positive feedback gains. The time derivative of (16) leads to:

$$\dot{\mu}(t) = (k_s + 1)r(t) + \beta_1 \operatorname{sign}[e_2(t)] \tag{17}$$

Since the nonlinearities in the system's dynamics are supposed to be unknown, a new control term, denoted $\hat{F}_d$, and generated by the NN feedforward estimation is added cancel out the effects of the uncertainties. $\hat{F}_d$ is then expressed by:

$$\dot{\hat{F}}_d = \hat{W}^T \sigma(V^T x_d) \tag{18}$$

where $V \in \mathbb{R}^{(a+1)\times L}$ is a bounded constant weight matrix, and $\hat{W} \in \mathbb{R}^{(L+1)\times 1}$, is the matrix of the estimates of the NN weights, generated on-line by:

$$\dot{\hat{W}} = K[\sigma(V^T x_d)e_2^T - \kappa\hat{W}] \tag{19}$$

where κ is a positive design constant parameter. $K = K^T > 0$ is a constant positive definit control gain matrix. According to property 1, the upper bound of $\dot{\hat{W}}$ can be formulated as follows:

$$\| \dot{\hat{W}} \| \leq F_N \sigma_m \| e_2 \| \tag{20}$$

where F_N is a known bound constant. The overall control input is then given by:

$$u = \hat{F}_d + \mu \tag{21}$$

By evaluating the time derivative of (21) and substituting the expressions of μ and $\dot{\hat{F}}_d$ given by (18) and (17) respectively, we get:

$$\dot{u} = \dot{\hat{F}}_d + \dot{\mu} = \hat{W}^T \sigma(V^T x_d) + (k_s + 1)r(t) + \beta_1 \operatorname{sign}[e_2(t)] \tag{22}$$

Thereby, the closed-loop tracking error system dynamics are formulated by considering the first time derivative of (10)

$$
\begin{aligned}
M(q)\dot{r} &= -\dot{M}(q)r + \dot{F}_d + \dot{S} - \dot{u} \\
&= -\dot{M}(q)r + \dot{F}_d + \dot{S} - \hat{W}^T \sigma(V^T x_d) - (k_s + 1)r(t) \\
&\quad -\beta_1 \operatorname{sign}[e_2(t)] \\
&= -\frac{1}{2}\dot{M}(q)r + \tilde{W}^T \sigma(V^T x_d) + \varepsilon(x_d) - (k_s + 1)r(t) \\
&\quad +(-\frac{1}{2}\dot{M}(q)r + \dot{S} + e_2) - \beta_1 \operatorname{sign}[e_2(t)] - e_2
\end{aligned} \tag{23}
$$

where $\tilde{W}^T = W^T - \hat{W}^T$ is the estimation error. Equation (23) can then be rewritten as follows:

$$M(q)\dot{r} = -\frac{1}{2}\dot{M}(q)r + \tilde{N} + N_{B_1} + N_{B_2} - e_2 - (k_s + 1)r(t) - \beta_1 \operatorname{sign}[e_2(t)] \tag{24}$$

with

$$\tilde{N} = -\frac{1}{2}\dot{M}(q)r + \dot{S} + e_2 \tag{25}$$

$$N_{B_1} = \varepsilon(x_d) \tag{26}$$

$$N_{B_2} = \tilde{W}^\top \sigma(V^\top x_d) \tag{27}$$

As detailed in [33], thanks to the Mean Value Theorem, $\tilde{N}$ is upper bounded as follows:

$$\| \tilde{N} \| = \| -\frac{1}{2}\dot{M}(q)r + \dot{S} + e_2 \| \leq \rho(\| z_1 \|) \| z_1 \| \tag{28}$$

where $z_1(t) \in \mathbb{R}^3$ is given by:

$$z_1(t) = \begin{bmatrix} e_1^\top & e_2^\top & r^\top \end{bmatrix}^\top \tag{29}$$

and $\rho(\| z_1 \|)$ is a positive non decreasing bounding function. In order to facilitate the stability analysis, some important inequalities are considered according to the following lemma.

Lemma 1
Consider N_{B_1} and N_{B_2} as expressed respectively by (26) and (27). The following inequalities hold.

$$\| N_{B_1} \| \leq \varepsilon_N \tag{30}$$

$$\| \dot{N}_{B_1} \| \leq \varepsilon_N' \tag{31}$$

$$\| N_{B_2} \| \leq (\bar{W}_m^\top + F_N \sigma_m \| e_2 \|)\sigma_m \equiv \xi_{B_2} \tag{32}$$

$$\| \dot{N}_{B_2} \| \leq \xi_1 \| e_2 \| + \xi_2 \tag{33}$$

where ξ_{B_2}, ξ_1, and ξ_2 are positive known constants.

Proof:
Inequalities (30) and (31) can be directly determined according to equations (14), (15), and (26). Based on Property 1 and equation (20) dealing with the upper bound of the NN weights, the inequality (32) can be easily justified. Then, by considering the derivative relation $\dot{\sigma}_m = \sigma_m(1 - \sigma_m)$ together with the time derivative of N_{B_2} expressed as $\dot{N}_{B_2} = \dot{\tilde{W}}\sigma_m + \tilde{W}\dot{\sigma}_m$, inequality (33) is concluded.

For a complete review of the stability analysis of the RISE-NN control approach, the reader is referred to [19].

5 Numerical Simulations

In this section, numerical simulations are conducted in the framework of Matlab/Simulink software. The 3.5" HDD (Seagate Barracuda 7200.10) dynamic model [34] is chosen as a demonstrator in simulation to test the effectiveness of the proposed control schemes. The full mathematical description of this prototype is given by equations (1)-(4) with a sample time fixed to $T_e = 0.05ms$. The normalized dynamic model parameters are given by: $M(q) = 1$, $\sigma_0 = 10^5$, $\sigma_1 = \sqrt{10^5}$, $\sigma_2 = 0.4$, $f_s = 1.5$,

$f_c = 1$, and $\dot{q}_s = 10^{-3}$. The NN weights are manually tuned off-line for the best possible controller performance. A physical restriction is imposed on the VCM actuator leading to the saturation of the control input u, that is $|u| \le 3v$ corresponding to a practical range in real HDD servo-systems. All initial conditions are chosen to be at the origin. Three different simulation scenarios have been considered.

The first scenario deals with tracking of both sinusoidal and step desired trajectory q_d without external disturbances [34]. The sinusoidal reference is chosen as $q_d = A \sin(\pi f t)$ where $A = 2\mu m$ and $f = 200Hz$, whereas the constant desired trajectory is chosen to be a unit step $q_d = 1\mu m$ and a zero mean value Gaussian white noise w_{noise} with a variance $\sigma^2 = 9 \times 10^{-9} (m)^2$ has been considered for this scenario as well as for the other scenarios.

In the second scenario, only the step response is investigated for a robustness test of the proposed controller towards external perturbations. Both input w_{in} and output w_{out} disturbances have been considered in this scenario, as schematically illustrated in Fig. 3. w_{out} is assumed to be an impulse disturbance with an amplitude of $0.3\mu m$ applied to the system at the time instance $t = 4ms$, while w_{in} is an unknown persistent bounded disturbance such that $w_{in} = -3mv$. In the third and last scenario, uncertainties on the system's inertia mass $M(q)$ are considered which can be caused by the movement of the R/W head tip onto the disc surface. Therefore, an efficient controller is required to ensure the convergence of the head to the target position, and to be robust enough against parametric uncertainties.

For all the proposed simulations scenarios described above, and in order to highlight the performance of the proposed control solution, the RISE-NN closed-loop responses are compared to those of a classical Proportional Derivative controller (PD). The PD control gains are manually tuned to get the best performance. However, automatic optimization tools can also be used to determine these parameters based on a suitable objective function.

For the purpose of performance comparison, an energy function E is introduced, it is expressed by:

$$E = \sum_{i=1}^{N_{sim}} |u_i|, \tag{34}$$

where N_{sim} is the sample number for the whole simulation duration.

Table 1 summarizes the parameters of the proposed controllers, while Tab. 2 gives a summary of the overall closed-loop system's performance with the different control solutions.

Tab. 1. Summary of the controllers' parameters.

Reference (μm)		RISE-NN				PD	
q_d	α_1	α_2	K_s	β_1	K_p	K_d	
$2\sin(200\pi\ t)$	3000	2900	1850	1	2×10^7	10^3	
1	1500	1500	1850	1	2×10^7	10^3	

Tab. 2. Controllers performances comparison.

Without disturbances (Sinusoidal Reference)		
	PD	RISE-NN
Settling time	2.85 ms	1.62 ms
Maximum overshoot	40%	25%
Control input $\mid u \mid$	3 v	3 v
Energy function (E)	$3.11\times 10^2 v$	$5.38\times 10^2 v$

Without disturbances (Step response)		
	PD	RISE-NN
Settling time	5.55ms	2.33 ms
Maximum overshoot	50%	1.5%
Control input $\mid u \mid$	3 v	1.07 v
Energy function (E)	$3.03\times 10^2 v$	43.44v

Disturbances Rejection		
	PD	RISE-NN
Recovery time	3.2 ms	2.7 ms
Maximum overshoot	9.5%	15%
Control input $\mid u \mid$	3v	1.5v
Energy function (E)	$3.84\times 10^2 v$	$1.2\times 10^2 v$

Parameters uncertainties (80% of error)		
	PD	RISE-NN
Settling time	9.8 ms	5.6 ms
Maximum overshoot	62%	15%
Control input $\mid u \mid$	3 v	1.5 v
Energy function (E)	4.83×10^2 v	77.66 v

5.1 Scenario 1: Tracking Problem in Nominal Case

Comparison between the performance of RISE-NN and PD controllers is illustrated in Figs. 5–9.

(a)

(b)

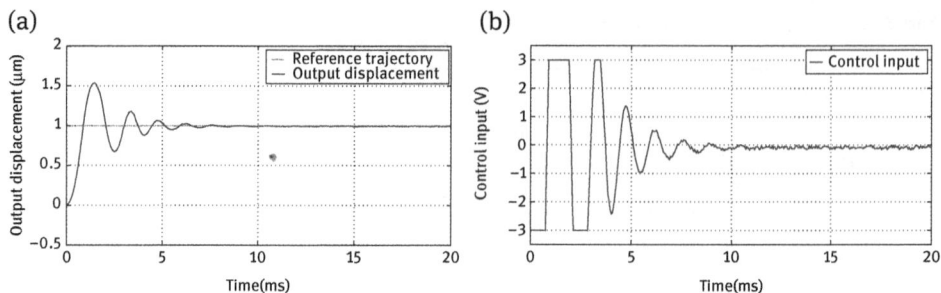

Fig. 5. Step response in non disturbed case with PD controller: (a) Output displacement and (b) Control input.

(a)

(b)

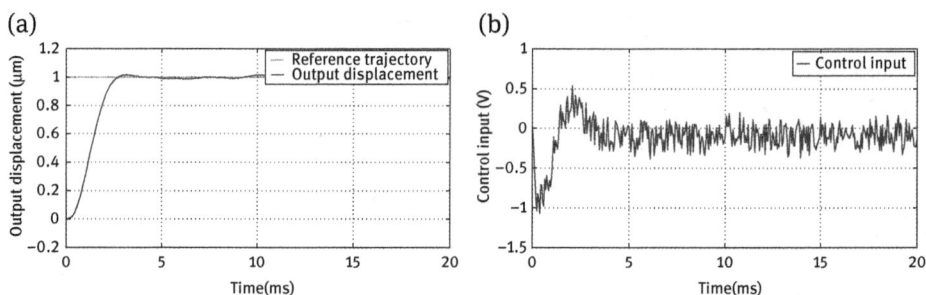

Fig. 6. Step response in non disturbed case with RISE-NN controller: (a) Output displacement and (b) Control input.

(a)

(b)

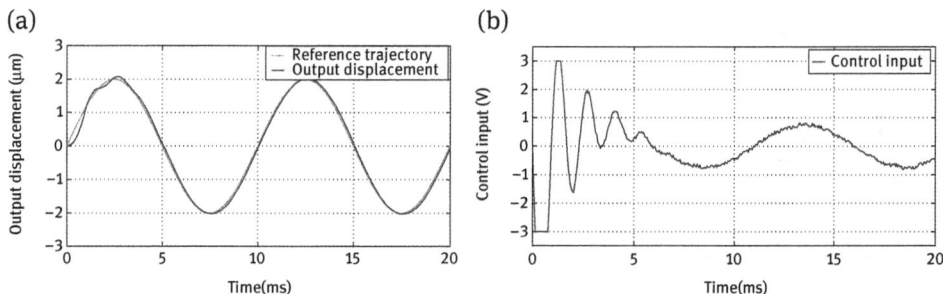

Fig. 7. Tracking of a sinusoidal reference trajectory in non disturbed case with PD controller: (a) Output displacement and (b) Control input.

First, in the case of the time varying reference signal (cf. Figs. 7–8), it can be observed that the proposed control solution RISE-NN achieves a better track following performance. Indeed, the head tip converges faster to the target position such that the tracking error is reduced around zero and little overshoots are generated.

(a)

(b)

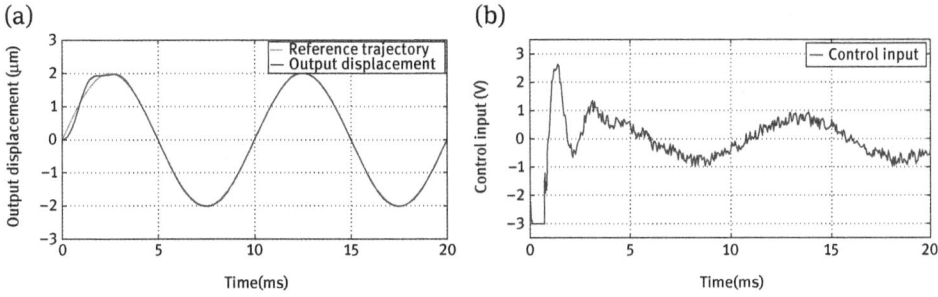

Fig. 8. Tracking of a sinusoidal reference trajectory in non disturbed case with RISE-NN controller: (a) Output displacement and (b) Control input.

(a)

(b)

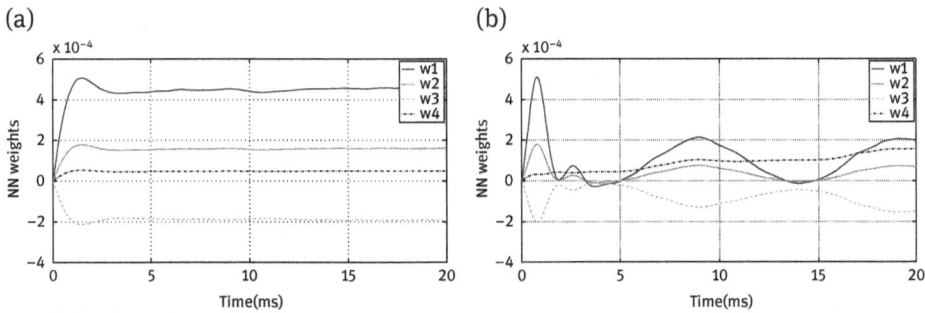

Fig. 9. Time history of the neural network weights in nominal case: (a) step response and (b) sinusoidal reference.

Second, the investigated step response (cf. Figs. 5–6) shows that RISE-NN tracking performances are much better than the PD controller. This is perceptible through the decreased overshoots which are negligible compared to those of the PD controller. In term of speed and precision, the RISE-NN settling time is substantially reduced, i.e. the R/W head tip reached the target and remains around such that the positioning accuracy is ensured.

In addition, the RISE-NN control energy consumption is very low in the case of step response, that is the control input is kept within the admissible limits.

The time history of the NN weights is displayed in Fig. 9 and shows that in both cases, sinusoidal and constant reference trajectories, the boundedness of the NN weight is ensured.

5.2 Scenario 2: External Disturbances Rejection

The main objective of this scenario is to test the effect of external disturbances on the closed-loop system and how the proposed control solutions deal with. The resulting disturbance rejection simulations with both PD and RISE-NN controllers are illustrated in Figs. 10, 11 respectively.

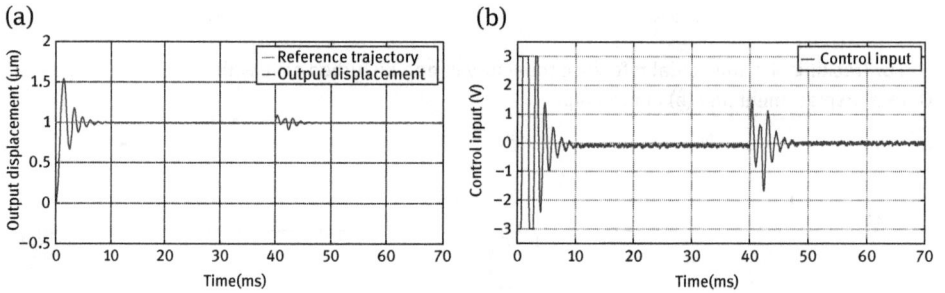

Fig. 10. Tracking under external disturbances with PD controller: (a) Output displacement and (b) Control input.

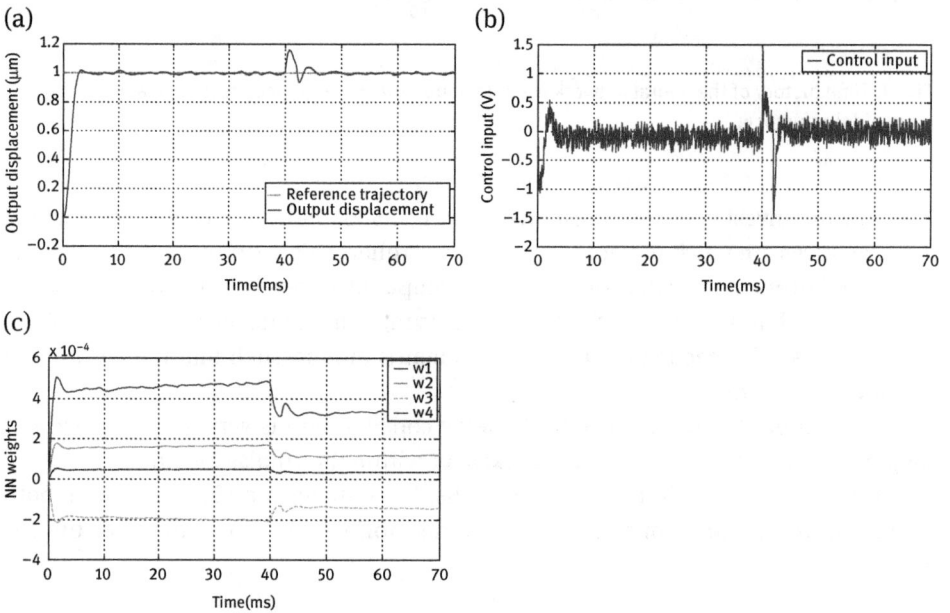

Fig. 11. Tracking under external disturbances with RISE-NN controller: (a) Output displacement, (b) Control input, and (c) Time history of the neural network weights.

For a persistent external disturbance ω_{in} and a punctual output disturbance ω_{out} applied at the time instant $t = 40ms$, both controllers successfully performed the rejection and the head tip is returned to its target track. However, with RISE-NN, the displacement of the head, as shown in Fig. 11(a), is faster with a reduced 5% settling time which results in a quick return of the position error signal to around zero, but the overshoot remains relatively significant.

Moreover, the RISE-NN control input evaluation satisfies the physical constraints and is kept within the interval $[-3v, 3v]$ without reaching the saturation limits. This results in a reduced control effort reflected by the energy function E. It is worth noting that the norm of the NN weights can be easily upper bounded by a constant as depicted in Fig. 11(c).

5.3 Scenario 3: Robustness Towards Parametric Uncertainties

For a complete comparative study between both proposed controllers, let us consider uncertainties on the system inertia of the HDD servo-system. The obtained simulation results for this case are shown in Figs. 12 and 13.

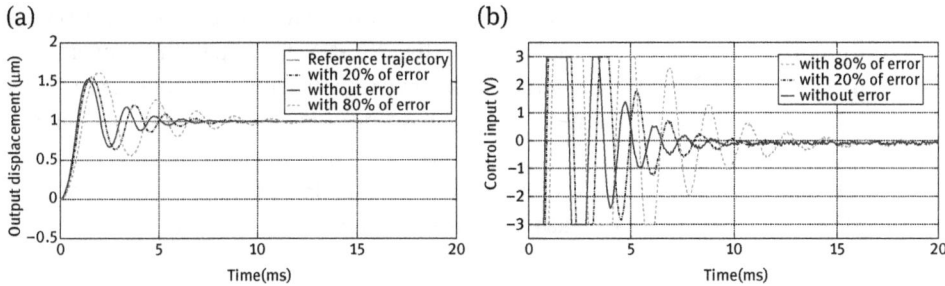

Fig. 12. Robustness towards parameters' uncertainties with PD controller: (a) Output displacement and (b) Control input.

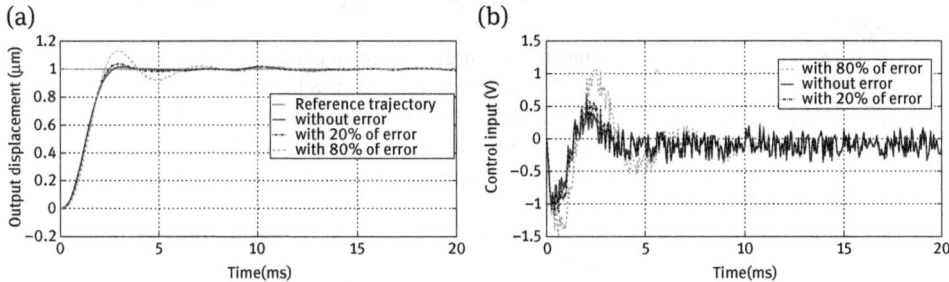

Fig. 13. Robustness towards parameters' uncertainties with RISE-NN controller: (a) Output displacement and (b) Control input.

For an uncertainty amount up to 80%, with respect to the nominal value, the closed-loop system behavior with a PD controller is so degraded. Therefore, as illustrated in Fig. 12, significant overshoots can be observed and the 5% settling time is too long.

However, with a RISE-NN controller, the system shows a good robustness against these uncertainties and the head tip achieves a quick convergence to the target position with a relatively smaller overshoots.

Consequently, according to these results, it can be concluded that against parametric uncertainties, the RISE-NN control solution is able to ensure better tracking performances.

6 Conclusion and Future Work

In this paper, a RISE based NN controller was proposed as a solution to solve a Hard-Disc-Drive head track following problem. The main objective was to guarantee that the tracking error is minimized as much as possible and converges to a neighborhood of zero.

Numerical simulations for different operating conditions are provided to show the effectiveness of the proposed controller. Thereby, through a comparative study with a PD controller, it was clearly shown that, with the RISE-NN approach, the convergence of the head tip to the target track can be ensured with superior performance in terms of speed, accuracy and robustness against external disturbances and parametric uncertainties.

Our future work will be focused on the optimization tools for an extended version of the RISE-NN controller as well as real-time experiments.

Bibliography

[1] B. Chen, K. Lee, T.H.and Peng and V. Venkataramanan. *Hard disc drive servo-systems*. Springer, 2006.
[2] B. Isayed and M. Hawwa. A nonlinear PID control scheme for hard disc drive servo-systems. *Mediterranean Conf. on Control and Automation*, MED'07, :1–6 June 2007.
[3] J. Ishikawa and M. Tomizuka. A novel add-on compensator for cancellation of pivot nonlinearities in hard disc drives. *IEEE Trans. on Magnetics*, 34(4):1895–1897, 1998.
[4] T. Atsumi, A. Okuyama, and M. Kobayashi. Track-following control using resonant filter in hard disc drives. *IEEE/ASME Trans. on Mechatronics*, 12(4):472–479, 2007.
[5] J. Chang and H. Ho. LQG/LTR frequency loop shaping to improve TMR budget. *IEEE Trans. on Magnetics*, 35(5):2280–2282, 1999.
[6] Z. Li, G. Guo, B. Chen and T. Lee. Optimal control design to achieve highest track-per-inch in hard disc drives. *J. of Information Storage and Processing Systems*, 3(1–2):27–41, 2001.

[7]   B. Chen, T. Lee, K. Peng, and V. Venkataramanan. Composite nonlinear feedback control for linear systems with input saturation: theory and an application. *IEEE Trans. on Automatic Control*, 48(3):427–439, 2003.

[8]   Z. Goh, T.B.and Li, B. Chen, T. H. Lee and T. Huang. Design and implementation of a hard disc drive servo system using robust and perfect tracking approach. *IEEE Trans. on Control Systems Technology*, 9(2):221–233, 2001.

[9]   R. Horowitz and B. Li. Adaptive control for disk file actuators. 34th *Conf. on Decision and Control*, 1:655–660, New Orleans, LA, 1995.

[10] M. Kobayashi, S. Nakagawa and H. Numasato. Adaptive control of dual-stage-actuator for hard disc drives. *American Control Conf.*, 1:523–528, Boston, MA, USA, 2004.

[11] T. Yamaguchi, K. Shishida, S. Tohyama, Y. Soyama, H. Hosokawa, H. Ohsawa, H. Numasato, T. Arai, K. Tsuneta and H. Hirai. A mode-switching controller with initial value compensation for hard disc drive servo control. *Control Engineering Practice*, 5(11):1525–1532, 1997.

[12] T. Yamaguchi and H. Hirai. Control of transient response on a servosystem using mode-switching control, and its application to magnetic disk drives. *Control Engineering Practice*, 6(9):1117–1123, 1998.

[13] J. Nie, E. Sheh, and R. Horowitz. Optimal H_∞ control for hard disc drives with an irregular sampling rate. *American Control Conf.*, CA, USA, 2011.

[14] M. Graham and R. De Callafon. Modeling and robust control for hard disc drives. *ASME/JSM Conf. on Micromechatronics for information and Precision Equipement (MIPE)*, Santa Carla, CA, USA, June 2006.

[15] M. Taktak, A. Chemori, J. Ghommam and N. Derbel. Track following control using nonlinear model predictive control in Hard disc drives. *Int. Conf. on Intelligent Robots and Systems*, 4401–4406, Tokyo, Japon, 2013.

[16] M. Taktak, A. Chemori, J. Ghommam and N. Derbel. Model predictive tracking control for a head-positioning in a hard disc drive. 21st *Mediterranean Conf. on Control and Automation*, 1368–1373, Platanias, Chania, Crete, Greece, June 2013.

[17] B. Xian, D. Dawson, M. De Queiroz and J. Chen. A continuous asymptotic tracking control strategy for uncertain nonlinear systems. *IEEE Trans. on Automatic Control*, 49(7):1206–1211, 2004.

[18] C. Makkar, G. Hu, W. Sawyer and W. Dixon. Lyapunov-based tracking control in the presence of uncertain nonlinear parameterizable friction. *IEEE Trans. on Automatic Control*, 52(10):1988–1994, 2007.

[19] P. M. Patre, W. MacKunis, K. Kent and W. Dixon. Asymptotic tracking for uncertain dynamic systems via a multilayer neural network feedforward and RISE feedback control structure. *IEEE Trans. on Automatic Control*, 53(9):2180–2185, 2008.

[20] D. Abramovitch, F. Wang and G. Franklin. Disk drive pivot nonlinearity modeling part I: Frequency domain. *American Control Conf.*, :2600–2603, Baltimore, Maryland, USA, 1994.

[21] F. Wang, T. Hurst, G. Abramovitch and D.and Franklin. Disk drive pivot nonlinearity modeling part II: time domain. *American Control Conf.*, :2604–2607, Baltimore, Maryland, USA, 1994.

[22] C. De Wit, H. Olsson, J. Astrom and P. Lischinsky. A new model for control of systems with friction. *IEEE Trans. on Automatic Control*, 40(3):419–425, 1995.

[23] B. Armstrong-Helouvry, P. Dupont and C. de Wit. A survey of analysis tools and compensation methods for the control of machines with friction. *Automatica*, 30(7):1083–1138, 1994.

[24] L. Gong, J.Q.and Guo, H. Lee and B. Yao. Modeling and cancellation of pivot nonlinearity in hard disk drives. *IEEE Trans. on Magnetics*, 38(5):560–3565, 2002.

[25] J. Swevers, F. Al-Bender, T. Ganseman and C.G. Prajogo. An integrated friction model structure with improved presliding behavior for accurate friction compensation. *IEEE Trans. on Automatic Control*, 45(4):675–686, 2000.

[26] J. Ishikawa and M. Tomizuka. Pivot friction compensation using an accelerometer and a disturbance observer for hard disk drives. *IEEE/ASME Trans. on Mechatronics*, 3(3):194–201, 1998.

[27] A. Ramasubramanian and L. E. Ray. Adaptive friction compensation using extended Kalman-Bucy filter friction estimation: a comparative study. *American Control Conf., 2588–2594*, Chicago, Illinois, June 2000.

[28] P. San, B. Ren, S. Ge and T. Lee. Adaptive neural network control of hard disk drives with hysteresis friction nonlinearity. *IEEE Trans. on Control Systems Technology*, 19(2):351–358, 2009.

[29] K. J. Astrom and C. De Wit. Revisiting the LuGre model, stick-slip motion and rate dependence. *IEEE Control Systems*, 28(6):101–114, 2008.

[30] D. Meddah, A. Benallegue and A. Cherif. A neural network robust controller for a class of nonlinear MIMO systems. *IEEE Int. Conf. on Robotics and Automation*, :2645–2650, Albuquerque, NM, 1997.

[31] F. Sun, L. Li and H. Li. Neuro-fuzzy dynamic-inversionbased adaptive control for robotic manipulatorsldiscrete time case. *IEEE Trans. on Industrial Electronics*, 54(3):1342–1351, 2007.

[32] T. Dierks and S. Jagannathan. Neural network control of mobile robot formations using RISE Feedback. *IEEE Trans. on Systems, Man and Cybernetics*, Part B, 39(2):332–347, 2008.

[33] P. M. Patre, W. MacKunis, K. Kent and W. Dixon. Asymptotic tracking for systems with structured and unstructured uncertainties. *IEEE Trans. on Control Systems Technology*, 16(2):373–379, 2008.

[34] C. De Wit, H. Olsson, J. Astrom and P. Lischinsky. Adaptive neural network control of hard disk drives with hysteresis friction nonlinearity. *IEEE Trans. on Control Systems Technology*, 19(2):351–358, 2011.

Biographies

Manel Taktak Meziou was born in 1984 in Sfax (Tunisia). She received her Engineer Degree in Electrical Engineering from the National School of Engineering of Sfax (Tunisia) in June 2008. Then, she obtained her Master Degree of Automatic Control and Industrial Computing from the National School of Engineering of Sfax in Tunisia in July 2009, and later in 2014 the PhD in Automatic Control Theory from the university of Sfax. Her research interests are in the area of nonlinear systems, control theory and robotic.

Ahmed Chemori received his MSc and PhD degrees respectively in 2001 and 2005, both in automatic control from the Grenoble Institute of Technology. He has been a Post-doctoral fellow with the Automatic control laboratory of Grenoble in 2006. He is currently a tenured research scientist in Automatic control and Robotics at the Montpelier Laboratory of Informatics, Robotics, and Microelectronics. His research interests include nonlinear, adaptive and predictive control and their applications in humanoid robotics, underactuated systems, parallel robots, and underwater vehicles.

Jawhar Ghommam was born in Tunis in 1979. He got the BSc degree in Computer and Control Engineering from the National Institute and Applied Sciences and Technology (INSAT) in 2003 in Tunis. He got the DEA (MSc) degree from the university of Montpelier at the Laboratoire d'Informatique, Robotique et Micro-électronique (LIRMM, France) in 2004, and later on in 2008 a Ph.D in Control Engineering degree from the university of Orleans. From 2008, he is with the National Institute of Applied Sciences and Technology, where he holds a tenured Assistant Professor at the Department of Electrical and Control Engineering. Dr. J. Ghommam occupies a permanent research position at the Research Unit on Intelligent Control and Optimization of Complex System (ICOS) at Sfax-Tunisia.

Nabil Derbel was born in Sfax (Tunisia) in April 1962. He received his Engineering Diploma from the National School of Engineering of Sfax in 1986, the DEA diploma in Automatics from INSA de Toulouse in 1986, the PhD. degree from the Laboratoire LAAS Of Toulouse in 1989, and the Doctorat dÉtat degree from the National School of Engineering of Tunis in 1997. He joined the Tunisian University in 1989, where he held different positions involved in research and education. Currently, he is Professor of Automatic Control at the National School of Engineering of Sfax. His current interests include optimal control, sliding mode control, fuzzy systems, neural networks.

I. Maalej, D. Ben Halima Abid, C. Rekik and N. Derbel

Single and Multiple Faults Detection in Nonlinear System Using Fuzzy Kalman Observer

Abstract: This paper deals with the problem of fault detection and state estimation for stochastic nonlinear systems. Cases of single, multiple and simultaneous actuator faults are considered. A fuzzy augmented state Kalman observer (FASKO) is developed to solve the problem stated above. The proposed observer is based on fuzzy logic and Kalman filter's theory. It is applied for Takagi-Sugeno fuzzy model which describes the nonlinear dynamic model by its decomposition into a number of linear submodels. Having these submodels, the Kalman equations are designed for each local model. This combination has been illustrated by comparing the performances of the FASKO with the classical augmented state Kalman observer(ASKO). Simulation results, performed on three tank system, validate the effectiveness of the proposed observer.

Keywords: Fault detection, state estimation, fuzzy augmented state Kalman observer.

1 Introduction

Interest in fault detection in nonlinear systems has grown significantly in recent years [2]. In fact, it is considered as one of the important step towards fault tolerant control systems (FTCs, [1], [3]). The main objective of fault detection is to provide early warnings to operators in order to take appropriate actions to prevent the break down of the system after the occurence of the faults. In the litterature, several methods are found for model based fault detection ([19],[20]). The most common method relies on the use of observers ([4],[5]) because they offers the opportunity to reconstruct the state variables of the process since the states are not always accessible and the transducers are not available or very expensive.

Luenberger observer was firstly proposed and applied in [6] for deterministic continuous time-invariant systems. Several studies were developed in order to include time-variant, discret and stochastic issues [7]. The problem of state estimation in stochastic linear system is solved by the well known Kalman filter. This filter [9], becomes more useful even for real-time application. An extension of the Kalman filter for nonlinear system was effectuated to get an Extended Kalman filter (EKF) [10]. In addition, to deal with the nonlinearity in the fault detection problem, Patton [18] has

I. Maalej, D. Ben Halima Abid, C. Rekik and N. Derbel: Control and Energy Management Laboratory (CEM Lab) , University of Sfax, Sfax Engineering School, Sfax, Tunisia, emails: imenmaalej@hotmail.fr donia.benhalimaabid@gmail.com, chokri.rekik@enis.rnu.tn, n.derbel@enis.rnu.tn.
De Gruyter Oldenbourg, ASSD – Advances in Systems, Signals and Devices, Volume 5, 2018, pp. 241–256.
DOI 10.1515/9783110470468-014

introduced the T-S fuzzy observer. The main principle is to use the T-S fuzzy models to represent nonlinear dynamic systems ([11],[12]). Then, the fuzzy observer is designed by a fusion of a number of local observers [16]. Authors in [14] and [15] have considered the design of fuzzy sliding mode observer for nonlinear systems which are subjected to unknown inputs or disturbances. A comparaison between observers based methods in model based fault detection was presented by Frank [22] in his survey paper. Later, Isermann [2] gave a survey on robust residual generation and evaluation methods used in observer based fault detection.

In this paper, the joint of fault and state estimation for a class of nonlinear systems are considered. Our main goal is to treat single, multiple and simultaneous actuator faults with their effects on the states. Therefore, an augmented state observer based on fuzzy logic and classical Kalman filter equations has been developed for noisy dynamic systems.

The remainder of this paper is organized as follows: In section 2, the problem to be considered is formulated. In section 3, the fuzzy augmented state observer using standard Kalman filter theory is presented. Section 4 is devoted to simulation results to illustrate the effectiveness of the proposed observer. Conclusions are drawn in section 5.

2 Problem Formulation

Consider the following linear time invariant state space model given by equation (1):

$$\begin{cases} X(k+1) = AX(k) + BU(k) + Ww(k) \\ Y(k) = CX(k) + v(k) \end{cases} \tag{1}$$

where $X(k) \in R^n$ is the state vector of the system, $U(k) \in R^m$ is the known system inputs and $Y(k)$ denotes the system outputs. Matrices A, B, C and W are known and have appropriate dimensions. $w(k)$ and $v(k)$ are process and measurement noises that are mutually uncorrelated:

$$E[w_i(k)v_j(k)^T] = 0$$

Both of the noises are assumed to be zero mean normally disturbed white noise sequences with covariance Q and R, respectively:

$$E[w_i(k)w_j(k)^T] = Q\delta_{ij} \; ; \; (Q \geq 0)$$
$$E[v_i(k)v_j(k)^T] = R\delta_{ij} \; ; \; (R \geq 0)$$

where T denotes transpose and δ_{ij} the kronecker delta function.

To define the stochastic process, the statistical proprieties (mean and variance) of the initial state should be specified.

mean: $E[X(0)] = \bar{X}(0)$

variance: $E[(X(0) - \bar{X}(0))(X(0) - \bar{X}(0))^T] = P^x(0)$

In this work, the initial state is a gaussian random variable and it is uncorrelated with the process and measurement noises $w(k)$ and $v(k)$:

$$E[X_0(k)w(k)^T] = 0; E[X_0(k)v(k)^T] = 0$$

In fault free case, the process is described as given in equation (1). Otherwise, because of a fault were happened at the actuator or the sensor , the system (1) evolves as follow:

$$\begin{cases} X(k+1) &= AX(k) + BU(k) + Ww(k) + F_a f(k) \\ Y(k) &= CX(k) + v(k) + F_s f(k) \end{cases} \tag{2}$$

where $f(k)$ is the fault vector. F_a is a matrix that signifies an actuator fault which affect the input signal. A sensor fault F_s provides a measurement error.

The initial fault satisfy the followings:

$E[f(0)] = \bar{f}(0)$

$E[(f(0) - \bar{f}(0))(f(0) - \bar{f}(0))^T] = P^f(0)$

The fault is generated by:

$$f(k+1) = f(k) \tag{3}$$

Treating $X(k)$ and $f(k)$ as an augmented state system by combining equations (2) and (3), we obtain the following representation:

$$\begin{cases} X_a(k+1) &= A_a X_a(k) + B_a U(k) + W_a w(k) \\ Y(k) &= C_a X_a(k) + v(k) \end{cases} \tag{4}$$

where:

$$X_a = \begin{bmatrix} X(k) \\ f(K) \end{bmatrix}, A_a = \begin{bmatrix} A & F_a \\ 0 & I \end{bmatrix}, B_a = \begin{bmatrix} B \\ 0 \end{bmatrix}; C_a = \begin{bmatrix} C \\ F_s \end{bmatrix}; W_a = \begin{bmatrix} W \\ 0 \end{bmatrix}$$

An alternative state space model for fault estimation which is based on the augmented state vector X_a is then introduced.

Consider that the actuator fault and the sensor fault didn't occur simultaneously. The matrice values of F_a and F_s are:

- In case of actuator fault: $F_a = B$; $F_s = 0$
- In case of sensor fault: $F_a = 0$; $F_s = I$

In this paper, we will adopt the augmented state space model as presented in eqution (4) in order to estimate the joint of fault and state.

3 Observer Structure

In this section, we will recall the Kalman filter's principle [8]. Then we combine its equations with fuzzy logic to develop a fuzzy augmented state Kalman observer (FASKO).

3.1 Augmented State Kalman Observer

The Kalman filter [9], is known to be an optimal filter for linear system. It has been widely used for stochastic estimation. In this section, we consider the augmented state Kalman observer (ASKO) for the joint of fault detection and state estimation.

The principle of the Kalman filter is determined by a recurrence relation based on two steps: prediction and correction.

$$\hat{x}(k|k) \rightarrow \hat{x}(k+1|k) \rightarrow \hat{x}(k+1|k+1)$$

where: $\hat{x}(k|k)$: the current state estimate

$\hat{x}(k+1|k)$: A priori state estimate (prediction step)

$\hat{x}(k+1|k+1)$: A posteriori state estimate (correction step)

In the predection step, the augmented state space model shown in equation (4) is used:

1. to predict the states at time instant $(k + 1)$ based on the estimated states at the previous time to get a priori estimation of the current state:

$$\widehat{X}_a(k+1|k) = A_a \widehat{X}_a(k) + B_a U(k) \tag{5}$$

2. to project the error covariance matrix to get a priori covariance matrix of the prediction error $e(k+1|k) = X_a(k+1) - \widehat{X}_a(k+1|k)$:

$$P_a(k+1|k) = A_a P_a(k|k)(A_a)^T + W_a Q(W_a)^T \tag{6}$$

$P_a(k|k) = E[e(K)e(K)^T]$ is the covariance matrix of the estimation error $(e(k|k) = X_a(k) - \widehat{X}_a(k|k))$ where the initial covariance matrix $P_a(0|0)$ is:

$$P_a(0|0) = \begin{bmatrix} P^x(0) & 0 \\ 0 & P^f(0) \end{bmatrix}$$

In the second step, the estimated state shown in equation 5 will be corrected. This step requires:

1. to compute the Kalman filter gain:

$$K_a(k+1) = P_a(k+1|k)(C_a)^T [C_a P_a(k+1|k)(C_a)^T + R]^{-1} \tag{7}$$

2. to correct the augmented state vector to get the posteriori state estimate:

$$\widehat{X}_a(k+1|k+1) = \widehat{X}_a(k+1|k) + K_a(k+1)[Y(k+1) - C_a\widehat{X}_a(k+1|k)] \qquad (8)$$

3. to update the estimation error covariance matrix to get the posteriori covariance matrix:

$$P_a(k+1|k+1) = [I - K_a(k+1)C_a]P_a(k+1|k) \qquad (9)$$

3.2 Fuzzy Augmented State Kalman Observer (FASKO)

In order to identify wether the actuator or the sensor is faulty, we will consider a fuzzy augmented state space model. The model is described by If-Then rules. Each rule represents local linear state space relations approximating the non linear systems. The ith rule of the Takagi-Sugeno model [16], [17], has the following forms:

Rule i : ($i = 1:N$, N is the number of the If-Then rules.)

If $z_1(k)$ is $M_{i,1}$ and...and $z_p(k)$ is $M_{i,p}$ Then

$$\begin{cases} X_i^a(k+1) &= A_i^a X_i^a(k) + B_i^a U(k) + W_i^a w(k) \\ Y_i(k) &= C_i^a X_i^a(k) + v(k) \end{cases} \qquad (10)$$

where $X^a(k) \in R^n$ is the state vector, $U(k) \in R^m$ is the input vector, $w(k) \in R^p$ is the process noise, $Y(k) \in R^q$ is the output vector and $v(k) \in R^n$ is the measurement noise. $z_1(k)...z_p(k)$ are the presmise variables and M_{ij} are the fuzzy sets. Matrices $A_i^a \in R^{n \times n}$, $B_i^a \in R^{n \times m}$ and $C_i^a \in R^{n \times q}$ denote respectively the augmented state matrix and the input matrix associated with the ith local model. The values of these matrices can be obtained after performing the direct linearization arround some operating points or by using identification procedure.

After defuzzification, we get the following equations:

$$\begin{cases} X^a(k+1) &= \sum_{i=1}^{N} h_i(z(k))X_i^a(k) \\ Y(k) &= \sum_{i=1}^{N} h_i(z(k))Y_i(k) \end{cases} \qquad (11)$$

The membership grade $h_i(z(k))$ are defined as:

$$h_i(z(k)) = \frac{w_i(z(k))}{\sum_{i=1}^{N} w_i(z(k))} \qquad (12)$$

$$w_i(z(k)) = \prod_{i=1}^{p} M_{ij}(z_j(k)) \qquad (13)$$

where $z(k) = [z_1(k)...z_p(k)]$. For all k, we have:

$$\begin{cases} \sum_{i=1}^{N} h_i(z(k)) = 1 \\ h_i(z(k)) \geq 0 \end{cases} \quad \text{and} \quad \begin{cases} \sum_{i=1}^{N} w_i(z(k)) \geq 0 \\ w_i(z(k)) \geq 0 \end{cases}$$

In reality, we cannot always observe all the state of a system. In this section, we propose fuzzy augmented state Kalman observer FASKO to estimate T-S model. It is assumed that the local linear models are uncorrelated and each of the N local model parameters is time invariant. At an operating point, Kalman filter which is associated with the ith fuzzy rule is presented as follows:

If $z_1(k)$ is $M_{i,1}$ and...and $z_p(k)$ is $M_{i,p}$ Then

$$\hat{x}_i^a(k+1|k) = A_i^a \hat{x}_i^a(k) + B_i^a U(k) \tag{14}$$

$$\hat{y}_i(k+1|k) = C_i^a \hat{x}_i^a(k+1|k) \tag{15}$$

$$P_i^a(k+1|k) = A_i^a P_i^a(k|k)(A_i^a)^T + W_i^a Q(W_i^a)^T \tag{16}$$

$$K_i^a(k+1) = P_i^a(k+1|k)(C_i^a)^T [C_i^a P_i^a(k+1|k)(C_i^a)^T + R]^{-1} \tag{17}$$

$$\hat{x}_a(k+1|k+1) = \hat{x}_a(k+1|k) + K_a(k+1)[Y(k+1) - C_a \hat{x}_a(k+1|k)] \tag{18}$$

$$P_i^a(k+1|k+1) = [I - K_i^a(k+1)C_a]P_i^a(k+1|k) \tag{19}$$

where K_i^a in equation (17) is the Kalman gain matrix and $P_i^a(k+1|k)$ and $P_i^a(k+1|k+1)$ are the covariance matrices of errors in predicted and updated state estimates of the ith local Kalman observer, respectively. The basic idea is that the Kalman observer in each fuzzy rule estimates the stats at any time instant in two steps (prediction and correction) which are explained in the previous section. The overall fuzzy augmented state Kalman observer is a weighted sum of individual local linear observer give by:

$$\begin{cases} \hat{X}_i^a(k+1|k+1) = \sum_{i=1}^{N} h_i(z(k))[\hat{x}_i^a(k+1|k) + K_i^a(k+1)[y_i(k+1) - \hat{y}_i(k+1|k)]] \\ \hat{Y}(k+1|k) = \sum_{i=1}^{N} h_i(z(k))C_i \hat{x}_i^a(k+1|k) \end{cases} \tag{20}$$

where the membership grade are defined as equations (12) and (13).

Fuzzy observers are required to satisfy $X^a(k) - \hat{X}^a(k) \to 0$ when $k \to \infty$. $X^a(k)$ and $\hat{X}^a(k)$ are the augmented state vector and the augmented state vector estimate, respectively. This condition guarantees that the steady error between $X^a(k)$ and $\hat{X}^a(k)$ converges to zero. But at first, we have to test the observability matrix for the obtained subsystems. A linear system is said to be observable if and only if the pair (A_i, C_i) satisfy the following equation:

$$\text{rank}(O) = n \quad \text{with} \quad O = [C_i C_i A_i C_i A_i^{n-1}] \tag{21}$$

4 Simulation Results

In this section, simulation studies on a three tank system [21] are achieved to validate the effectiveness of ASKO and FASKO algorithms.

4.1 System Description

Figure 1 shows a schematic diagram of a three tank system [21].

It consist of three cylindrical tanks T1,T2 and T3 wich are serially connected by cylindrical pipes with cross sectional area S_p.

Flow rates Q_{e1} of the first tank and Q_{e2} of the third tank are considered as process inputs.The water levels h_1, h_2 and h_3 represent the measurable outputs.

All technical data are given in Tab. 1.

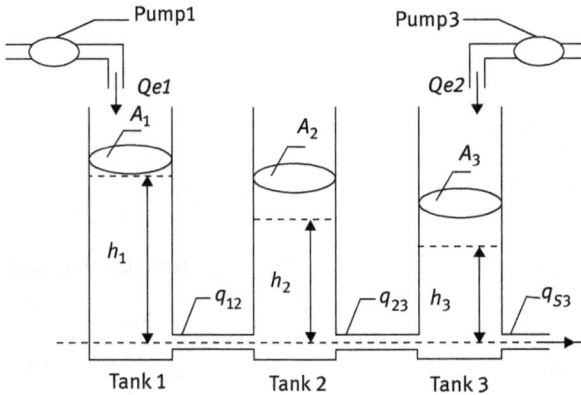

Fig. 1. Three tank system.

Tab. 1. System Parameters.

Parameters	Values
water level (h_1, h_2, h_3) (m)	$0.5, 0.45, 0.4$
max flow rate (Q_{e1}, Q_{e2}) (m³/sec)	5×10^{-5}
cross section area of the tank (A_1, A_2, A_3) (m²)	0.0154
cross section area of the connecting pipes (S_p) (m²)	5×10^{-5}
constant outflow coefficients (c_1, c_2, c_3)	$1, 0.8, 1$

After establishing the balance of inflow and outflow of each tank, we obtain the following non linear model:

$$
\begin{cases}
\dfrac{dh_1}{dt} = Q_{e1}(t) - q_{12}(t) \\
\dfrac{dh_2}{dt} = q_{12}(t) - q_{23}(t) \\
\dfrac{dh_3}{dt} = q_{23}(t) + Q_{e2}(t) - q_{s3}(t)
\end{cases}
\tag{22}
$$

where q_{ij} represents the water flow rate from tank i to tank j. By using the Bernouilli's law, the flow rate q_{ij} is expressed by equation (25):

$$
q_{ij} = c_i S_p \sqrt{2g\Delta h}
\tag{23}
$$

Put

$$
\frac{dh}{dt} = f(h, Q_e)
\tag{24}
$$

where

$$
f(h, Q_e) =
\begin{vmatrix}
\dfrac{1}{A_1}\left[Q_{e1} - c_1 S_p \sqrt{2g(h_1 - h_2)}\right] \\
\dfrac{1}{A_2}\left[c_1 S_p \sqrt{2g(h_1 - h_2)} - c_2 S_p \sqrt{2g(h_2 - h_3)}\right] \\
\dfrac{1}{A_3}\left[c_2 S_p \sqrt{2g(h_2 - h_3)} + Q_{e2} - c_3 S_p \sqrt{2g(h_3)}\right]
\end{vmatrix}
\tag{25}
$$

To put the equation (26) in a state space form, A_i and B_i are considered as the jacobian matrix of f;

$$
A_i = \left.\frac{\partial f}{\partial h}\right|_{h(t)=h_i} ;
\tag{26}
$$

$$
B_i = \left.\frac{\partial f}{\partial Q_e}\right|_{u(t)=Q_{ei}}
\tag{27}
$$

With the sampling time T= 5 s, simulation runs are performed using ASKO and FASKO algorithms for state and fault estimation.

According to ASKO algorithm, matrices A, B and C are obtained after performing linearization arround the nominal value shown in Tab. 1 and then, their discretisation.

The flow rates Q_{e1} and Q_{e2} are chosen equal to $2\ 10^{-5}\ m^3/sec$ and $3.375\ 10^{-5}\ m^3/sec$ respectively. The initial conditions in the simulations are assumed to be:

$$
\begin{bmatrix} h_1(0) \\ h_2(0) \\ h_3(0) \end{bmatrix} =
\begin{bmatrix} 0.4 \\ 0.35 \\ 0.3 \end{bmatrix} ;
\begin{bmatrix} \hat{h}_1(0) \\ \hat{h}_2(0) \\ \hat{h}_3(0) \end{bmatrix} =
\begin{bmatrix} 0 \\ 0 \\ 0 \end{bmatrix}
$$

We added normally disturbed noises w(k) and v(k) to the states and measurement outputs, respectively. The process noise is selected as 0.75% of the nominal inflow to tank 1 and 3. The measurement noise is selected as 0.25% of the normal height of the tanks.

For the FASKO, we present the system with T-S model. We propose to linearize the nonlinear model around some operating points. Nine local fuzzy augmented state Kalman observers are designed for state and fault estimation.

We can easily checked that the necessary conditions of the observability are satisfied.

Q_{e1} and Q_{e2} are the premise variables and the membership functions for these premise variables are given in Figs. 2(a) and 2(b).

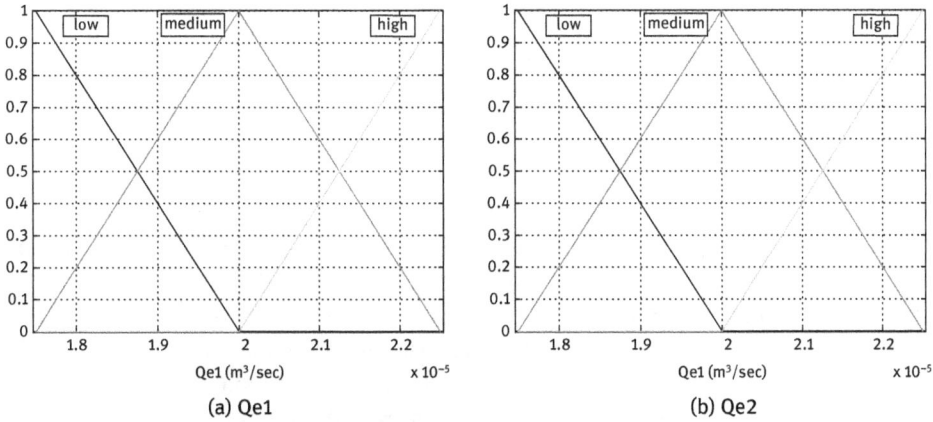

Fig. 2. Membership functions of the flow rates (a) Q_{e1} and (b) Q_{e2}.

4.2 Actuators Simulation Results

The three tank system is actuated by two pumps (two water flow rates Q_{e1} and Q_{e2}). In order to detect fault which can happen in one or two actuators, single,multiple ans simultaneous faults have been created on the system. It is assumed that the sensors are faultless (Matrix $F_s = 0$). In this paper, we treat only multiple and simultaneous actuator faults. The results of single fault case have been presented in details in [13].

The performance of ASKO and FASKO are analyzed after introduced the faults.

4.2.1 Multiple Sequential Faults

The actuator faults are applied as a ramp having zero initial value and appears at 500th sampling instant and 2000th sampling instant in actuator 1 and actuator 2, respectively.

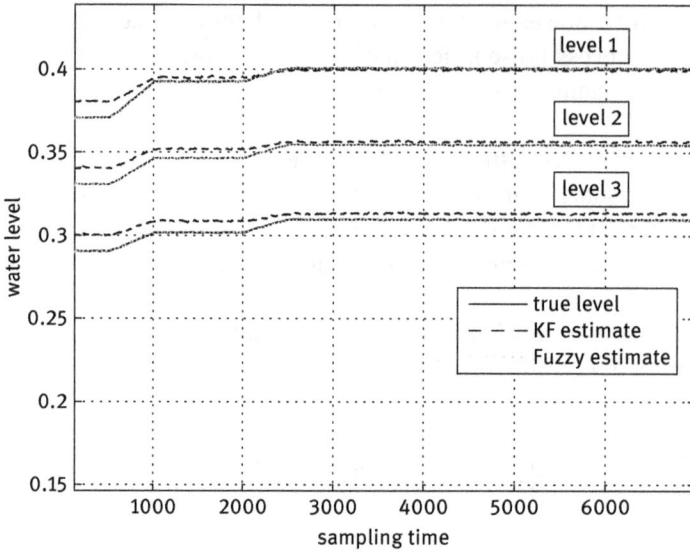

Fig. 3. Evolution of water level in case of multiple sequential faults.

The magnitude of the first actuator reached 0.75×10^{-6} m^3/sec at 1000th and the magnitude of the second actuator reached 4×10^{-6} m^3/sec at 2500th sampling instant. The two actuators faults remain constant up to 7000th sampling instant.

The state estimates given by ASKO and FASKO are presented in Fig. 3.

Figure 3 shows the evolution of true (h_1, h_2, h_3) and estimated ($\hat{h}_1$, $\hat{h}_2$, $\hat{h}_3$) water levels using ASKO and FASKO. Their estimations reveal an occurrence of faults until 2500th sampling instant. It is obvious that the fuzzy level estimates outperform the Kalman level estimates when multiple sequential faults have been happened. In fault free case, the level estimates are reasonably accurate for both of observers. However the magnitude of the faults still unknown. Therefore, we will based on the steady errors between the faults and their estimates.

Figures 4 and 5 present respectively the evolution of the true actuator faults and their estimates, and the evolution of the estimation errors between the true actuator faults and their estimates using ASKO and FASKO in case of multiple sequential faults.

The multiple sequential fault estimates shown in Fig. 4 are reasonably accurate and they are unbiased. Although the two actuator fault estimates are acceptable, the convergence of fuzzy fault estimates are faster than Kalman fault estimates.

In figures 5(a) and 5(b), two large peaks are occurred in two different time intervals. Beyond the two peaks, the estimation errors are close to zero in the fault free case. From these figures, we can confirm that the use of FASKO to estimate the errors provide more satisfactory results than ASKF.

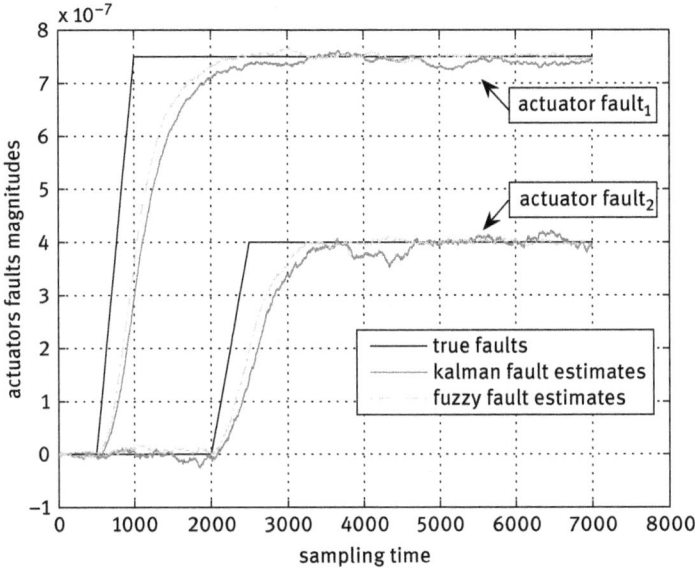

Fig. 4. Evolution of the multiple sequential faults.

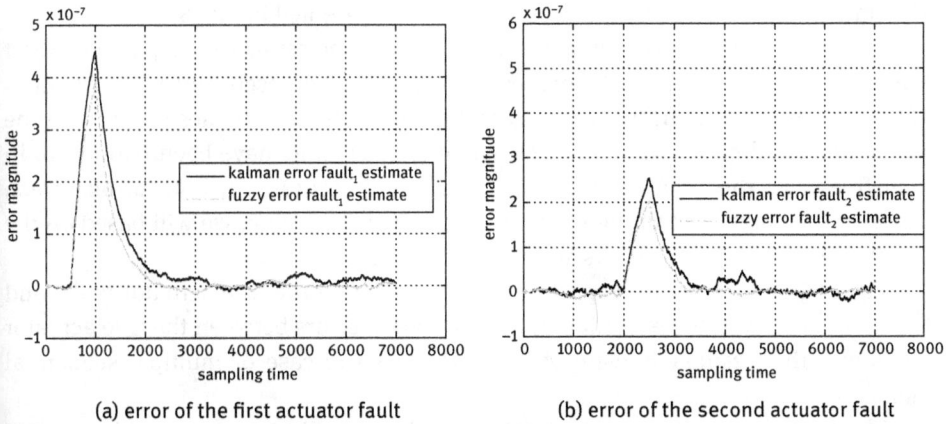

(a) error of the first actuator fault

(b) error of the second actuator fault

Fig. 5. Evolution of estimation error for the multiple sequential faults.

4.2.2 Multiple Simultaneous Faults

The two actuator faults are also applied as a ramp having zero initial condition. They appears at the same time at 500th sampling instant and their magnitudes reached $0.75e - 6$ m^3/sec and $0.4e - 6$ m^3/sec at 1000th sampling instant, respectively. The two actuators faults remain constant up to 7000th sampling instant.

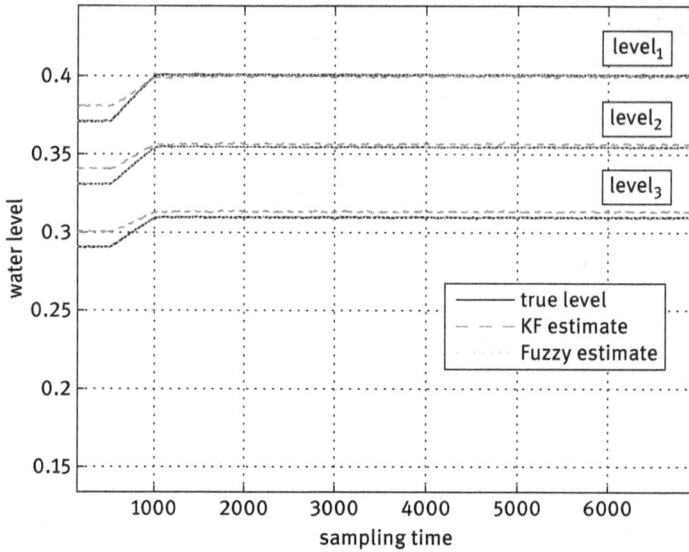

Fig. 6. Evolution of water level in case of multiple simultaneous faults.

The state estimates given by ASKO and FASKO are presented in Fig. 6.

Figure 6 shows the evolution of true (h_1, h_2, h_3) and estimated ($\hat{h}_1$, $\hat{h}_2$, $\hat{h}_3$) water levels using ASKO and FASKO. Their estimations reveal an occurrence of faults until 1000th sampling instant. It is obvious that the fuzzy level estimates outperform the Kalman level estimates when multiple simultaneous faults have been happened. In fault free case, the level estimates are reasonably accurate for both of observers. However the magnitude of the faults still unknown. Therefore, we will based on the steady errors between the faults and their estimates.

Figure 7 and 8 present respectively the evolution of the true actuator faults and their estimates, and the evolution of the estimation errors between the true actuator faults and their estimates using ASKO and FASKO in case of multiple sequential faults.

The multiple simultaneous fault estimates shown in Fig. 7 are reasonably accurate and they are unbiased. Although the two actuator fault estimates are acceptable, the convergence of fuzzy fault estimates are faster than Kalman fault estimates.

In figures 8(a) and 8(b), two large peaks are occurred in the same time interval. Beyond the two peaks, the estimation errors are close to zero in the fault free case. From these figures, we can confirm that the use of FASKO to estimate the errors provide more satisfactory results than ASKF.

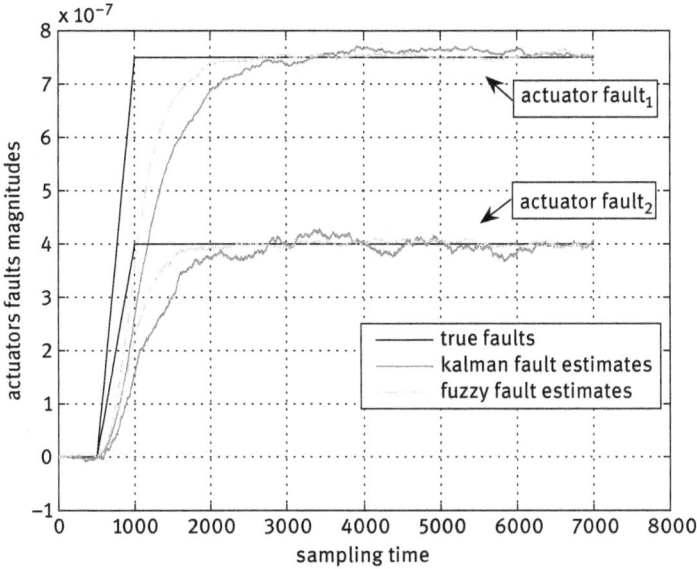

Fig. 7. Evolution of the multiple simultaneous faults.

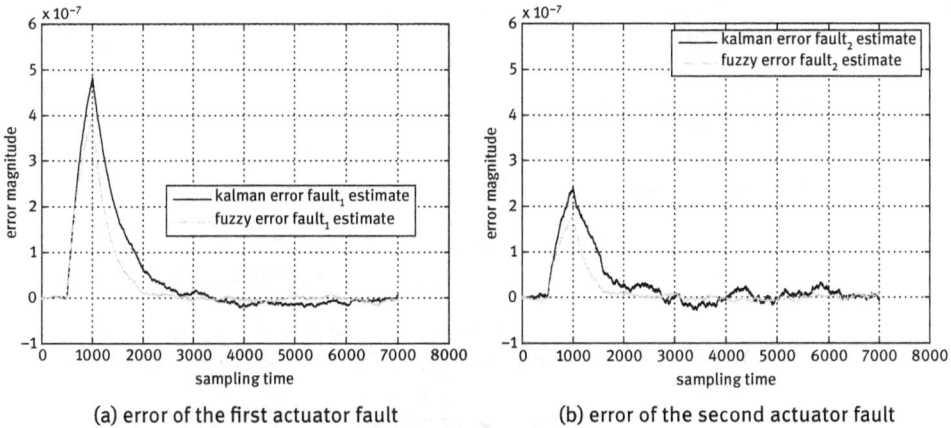

(a) error of the first actuator fault

(b) error of the second actuator fault

Fig. 8. Evolution of estimation error for the multiple sequential faults.

5 Conclusion

In this paper, we have treated the problem of fault detection and state estimation. The cases of more than one fault which can occur on the system actuator were studied. First at all, the problem is formulated as an augmented state space model. A fuzzy

Kalman observer has been developed for nonlinear systems which are represented by T-S fuzzy model to generate accurate state estimates in the presence of process and measurement noises. The proposed observer (FASKO) and the standard Kalman observer(ASKO) have been compared on the quality of state and fault estimation for a three tank system. The simulation results for both observers are quite satisfactory to detect multiple and simultaneous actuator fault. In addition, we infer that the FASKO presents a good alternative for the ASKO. However, the proposed observer converges more faster and seems to be less sensible to the noise than Kalman observer.

Bibliography

[1] R. Isermann. *Fault-diagnosis Systems an introduction from fault detection to fault tolerance.* Berlin, Germany, Springer-Verlag, 2006.
[2] R. Isermann. Model-based Fault-detection and Diagnosis - Status and Application. *Annual Reviews in Control*, 29:71–85, 2005
[3] R.J. Patton. Fault-tolerant control: the 1997 situation. *IFAC Symp. SAFEPROCESS'97: Fault detection, supervision, and safety for technical processes*, 2:1033–1055, Hull, UK, 1997.
[4] P. M. Frank and X. Ding. Survey of robust residual generation and evaluation methods in observer-based fault detection system. *J. of Process Control*, 7(6):403–424, 1997.
[5] H. B. Wang, J. L. Wang and J. Lam. Robust fault detection observer design: iterative LMI Approache. *ASME Journal of Dynamic Systems, Measurement and Control*, 129:77–82, 2007.
[6] D.G. Luenberger. Observing the state of a linear system. *IEEE Trans. on Military Electronics*, 8(2):74–80, 1964.
[7] D.G. Luenberger. An Introduction To Observers. *IEEE Trans. on Automatic Control*, 16(6):596–602, 1971.
[8] I.S.-H.T-Sai, M.-H.Lin, C.-H.Zheng, S.-M.Guo and L.-S.Shieh. Actuator fault detection and performance recovery with Kalman filter based adaptive observer. *Int. J. of General System*, 36:375–398, 2007.
[9] G.B. Greg Welch. *An Introduction To The Kalman Filter*. Copyright 2001 by ACM, Inc, 2001.
[10] G. Jonsson and O.P. Palsson. An application of extended Kalman filtering to heat exchanger models. *ASME*, 116:257–264, 1994.
[11] D. Ichalal, B. Marx, J. Ragot and D. Maquin. State estimation of nonlinear systems using multiple model approach. *American Control Conf.*, ACC, St. Louis, Missouri, USA, June 10–12, 2009.
[12] A. Khedher, K. Benothman, D. Maquin andM. Benrejeb. An approach of faults estimation in Takagi-Sugeno fuzzy systems. 8th *ACSIIEEE Int. Conf. on Computer Systems and Applications*, AICCS 2010, Hammamet, Tunisia, May 2010.
[13] I.Maalej, D. Ben halima, C.Rekik and N.Derbel. Fuzzy Augmented State Kalman Observer for Fault And State Estimation. In 11th *Int. Multi-Conf. on Systems, Signals and Devices*, Barcelona, Spain, February 11–14, 2014.
[14] R. Palm and P. Bergsten. Sliding mode observers for Takagi-Sugeno Fuzzy Systems. 9th *IEEE Int. Conf. on Fuzzy Systems*, FUZ'IEEE, 2000.
[15] A. Akhenak, M.Chadli, J. Ragot, and D. Maquin .Sliding mode multiple observer for fault detection and isolation. 42th *IEEE Conf. on Decision and Control*, Hawaii, December 9–12, 2003.

6] P. Bergsten, R. Palm and D. Driankov. Observers for Takagi-Sugeno Fuzzy Systems. *IEEE Trans. on Systems, Man and Cybernetics, Part B*, 32(1):114–121, February 2002.

7] D. Ichalal, B. Marx, J. Ragot and D. Maquin. Design of Observers for Takagi-Sugeno Systems with Immeasurable Premise Variables: an L2 Approach. 17th *World Congress, The Int. Federation of Automatic Control*, Seoul, Korea, July 6–11, 2008.

8] R. J. Patton, J. Chen and C. J. Lopez-Toribio. Fuzzy observer for nonliear dynamic systems. 37th *IEEE Conf. on Decision and Control*, Tampa Florida, 1:84–89, (CDC 98).

9] D.Fragkoulis, G.Roux, B.Dahhou. A global scheme for multiple and simultaneous faults in system actuators and sensors. 6th *Int. Multi-Conference on Systems, Signals and Devices*, Djerba, Tunisie, March 23–26, 2009.

20] D.Fragkoulis, G.Roux and B.Dahhou. Sensor fault detection and isolation observer based method for single, multiple and simultaneous faults: application to a waste water treatment process. *Int. Symp. on Advanced Control of Chemical Processes*, Istanbul, Turquie , July 12–15, 2009.

21] S.Abraham Lincon, D.Sivakumar and J.Prakash. State and Fault Parameter Estimation Applied To Three-Tank Bench Mark Relying On Augmented State Kalman Filter. *J. of Automatic Control and System Engineering (ICGST-ACSE)*, 7(1), May 2007

22] P. M. Frank. Fault diagnosis in dynamic system using analytical and knowledge based redundancy, a survey and some new result. *Automatica*, 26(3):459–474, 1990.

Biographies

Imen Maalej was born in Sfax, Tunisia, on July 1986. She received her engeneering diploma in industrial and automatic computing from National Engineering School of Sfax (ENIS), Sfax University, in June 2010. She received her master's diploma in the same field in June 2011. She is currently pursuing her PhD in the field of fault detection and isolation and fault tolerant control system. Her research interests include particulary process modeling, observers for nonlinear systems and process control.

Donia Ben Halima Abid received her National Engineering Diploma in electric and Automatic Engineering in 2000, her Master degree in Industrial and Automatic Computing in 2001 and her PhD in 2009 from the National School of Engineers of Sfax (ENIS), Tunisia. Currently, she is an assistant professor at the National School of Engineers of Sfax and she is teaching Linear Analysis and Control, Logic Systems, Nonlinear Analysis and Control, Robotic Modeling and Control, discrete analysis and Contro. Dr. Ben Halima is a Member of the research laboratory on Control and Energy Management. Her specific research interests deals with Adaptive Fuzzy Control, Fuzzy modeling and Identification.

Chokri Rekik was born in Sfax (Tunisia) on August 1974. He received the Engineering Diploma in 1999, the Diplme d'Etudes Approfondies in Automatic control in 2001 and a PhD degree in 2006 all from the Ecole Nationale d'Ingénieurs de Sfax (ENIS). He has been an Assistant Professor in the Electrical Engineering Department of the Ecole Nationale d'Ingénieurs de Sfax i 2006. Currently,he is a full Professor of Automatic Control at the Ecole Nationale d'Ingénieurs de Sfax and carrying his research activities in Contre and Energy Management Lab (CEM Lab) in ENIS.His research interests is: advanced control of complex nonlinear systems, robotic control.

Nabil Derbel was born in Sfax (Tunisia) in April 1962. He received the engineering Diploma from the Ecole Nationale d'Ingénieurs de Sfax in 1986 the Diplôme d'Etudes Approfondies in Automatic control from the Institut National des Sciences Appliquées de Toulouse in 1986, the Doctorat d'Université degree from the Laboratoire d'Automatique et d'Analyse des Systèmes de Toulouse in 1989, and the Doctorat d'Etat degree from the Ecol Nationale d'Ingénieurs de Tunis. He joined the Tunisian University in 1989, where he held different position involved in research and education. Currently, he is a full Professor of Automatic Control at the Ecole Nationale d'Ingénieurs de Sfax. He is an IEEE Senior member. His current interests include: Optimal Control, Complex Systems, Fuzzy Logic, Neural Networks, Genetic Algorithm. He is the author and the co-author of more than 50 papers published in international Journals and of more than 300 papers published in international conferences.